TRANSFORMING EDUCATION WITH NEW MEDIA

Studies in the Postmodern Theory of Education

Shirley R. Steinberg
General Editor

Vol. 435

The Counterpoints series is part of the Peter Lang Education list.
Every volume is peer reviewed and meets
the highest quality standards for content and production.

PETER LANG
New York • Washington, D.C./Baltimore • Bern
Frankfurt • Berlin • Brussels • Vienna • Oxford

Peter DePietro

TRANSFORMING EDUCATION WITH NEW MEDIA

Participatory Pedagogy, Interactive Learning, and Web 2.0

PETER LANG
New York • Washington, D.C./Baltimore • Bern
Frankfurt • Berlin • Brussels • Vienna • Oxford

Library of Congress Cataloging-in-Publication Data

DePietro, Peter.
Transforming education with new media: participatory pedagogy,
interactive learning, and Web 2.0 / Peter DePietro.
pages cm. — (Counterpoints. Studies in the postmodern theory of education; v. 435)
Includes bibliographical references.
1. Internet in education. 2. Internet in education—Case studies. 3. Online social
networks. 4. Online social networks—Case studies. I. Title.
LB1044.87.D46 004.67'8071—dc23 2012007810
ISBN 978-1-4331-1793-0 (hardcover)
ISBN 978-1-4331-1794-7 (paperback)
ISBN 978-1-4539-0831-0 (e-book)
ISSN 1058-1634

Bibliographic information published by **Die Deutsche Nationalbibliothek**.
Die Deutsche Nationalbibliothek lists this publication in the "Deutsche
Nationalbibliografie"; detailed bibliographic data is available
on the Internet at http://dnb.d-nb.de/.

The paper in this book meets the guidelines for permanence and durability
of the Committee on Production Guidelines for Book Longevity
of the Council of Library Resources.

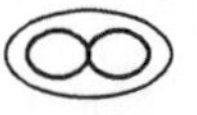

29 Broadway, 18th floor, New York, NY 10006
www.peterlang.com

Printed in the United States of America

INTRODUCTION

The possibilities that online platforms and new media technologies provide, in terms of human connection and the dissemination of information, are seemingly endless. With Web 2.0 there is an exchange of messages, visions, facts, fictions, contemplations, accusations, exclamations, and declarations buzzing around a network of computers that connects students to the world, fast. Theoretically this digital connectivity, and the availability of information that results from it, is beneficial to curriculum development in higher education. Education is easily available, democratic, and immersive. But is it worthwhile? Is the kind of education you can get from new media platforms and social media resources—with their click-on videos, rollover animations, and unfiltered content—of a quality that educators should be quick to integrate these tools into teaching?

This book examines the use of new media in pedagogy, as it presents case studies of the integration of online tools and social media in courses I, a professor of new media at an urban research university in the United States, teach or have taught. There is an assessment of the pedagogic endeavor in terms of benefits and risks to students, an analysis of interactive learning as it pertains to each case study, and an investigation into the future potential of new media concepts and technologies in higher education teaching. Technology can transform the process of education. However, we educators need to create standards that will guide students in the appropriate and responsible use of these tools. That way, education with technology-based models for teaching will produce meaningful learning.

1

WEB 2.0 AND NEW EDUCATION

For the purpose of this book, I am broadly defining Web 2.0 as a movement focused on technologies that engage the user. Typically these technologies are available to users via the Internet and on mobile devices. User-centered design, which involves users in an intuitive way, is necessary. More broadly, Web 2.0 is part of new media.

New media, contrasted with conventional (or old) media, rely on a digital signal instead of an analog signal to communicate message. New media include websites, wikis, interactive forums, e-learning systems, software, hardware, mobile devices, and the list goes on. Many of these items are discussed in this book. When successfully integrated into classroom instruction, new media can facilitate learning. A main reason for this is that technology spurs communication, supports information sharing, and creates access to data around the clock. In education there is an environment focused on technology use and integration that has become commonplace. Moreover, new media and associated Web 2.0 technologies can provide a foundation for the development of a new kind of education, one that is participatory and relies on digital innovation in the classroom. This *new education* engages students with technology like never before, and it provides unique and effective ways for students to interact with information. Thus, students learn differently. Web 2.0 can produce Education 2.0.

These are lofty terms: Web 2.0 and Education 2.0. But together they describe our advancement to a new level in education, made possible by new technologies—technologies that serve curriculum development and integrate well into teaching. Admittedly, in the age of new media, we tend to describe systems, methodologies, and processes that employ technology or that have technology as an integral component, with made-up, sometimes superfluous terminology. Here is a list of some examples: asynchronous interaction, blogosphere, social intermediary, mash-up, social news aggregator, and WYSIWYG. However, with a term like Education 2.0, we appropriately stress the importance of moving to a new level: the second level. It seems that in the history of education, advancing to a second level because of new media was inevitable. Describing the transformation with the label Education 2.0 seems right. With this label, we reference the technology-focused concepts of Web 2.0 and give importance to our use of new media in the classroom. New media in education is transformative, and it is the foundation of new education—education at this new level.

It is important to acknowledge that new media is an ever-changing discipline. More than a discipline, new media is a bundle of concepts, a set of technologies, and a trend. New media is also a reinvention of older methods of content production and delivery, just now digital. It is all of these things—discipline, technologies, concept, trend, and method for content production and delivery—and it is *new*. A question must be asked: What is *new* about new media, and is it *always* new?

The tools, technologies, and platforms of new media become obsolete quickly. New versions of hardware and software seem to be in an endless cycle of reinvention.

Technology ages so quickly that many institutions update computer labs, in terms of both hardware and software, when budgets allow, not when new versions come out. Online learning systems like Blackboard become digital dinosaurs before instructors and students can master updates. Social networking platforms that are integrated into course instruction may disappear. These networks may be bought by a competitor or be losers in an ongoing battle for market share. Because the tools, technologies, and platforms change, the theories educators integrate into teaching new media must also change. There needs to be a theoretical dynamism that matches the ongoing evolution of the tools. This combined theoretical and practical flux is real. It is necessary to embrace this *constant inconstancy* when educating with new media.

With new media, a successful pedagogy is an evolutionary one. This, perhaps, is the biggest challenge in new education. The concept of an evolu-

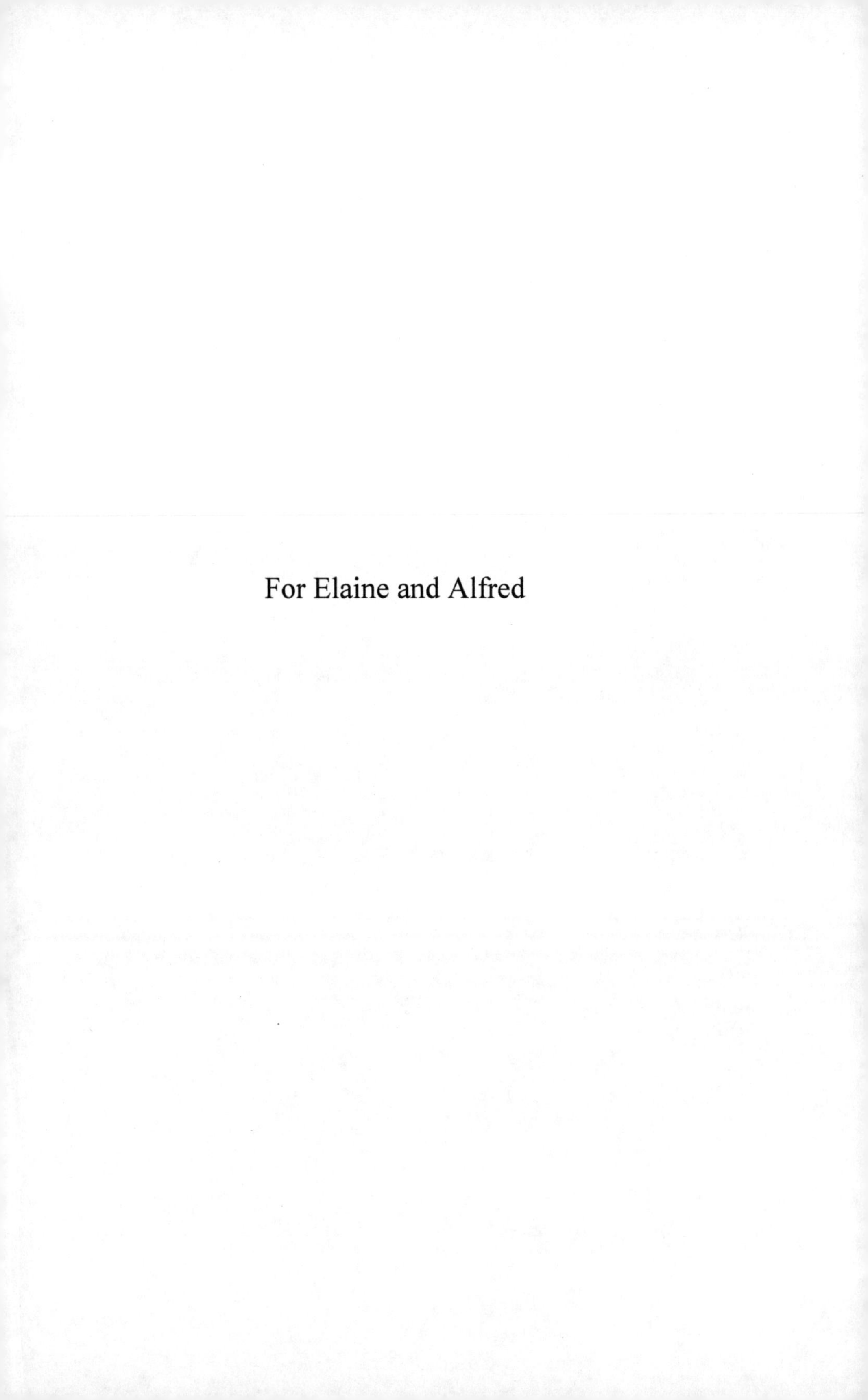

For Elaine and Alfred

CONTENTS

tionary pedagogy is relatively new. In the age of new media, as pedagogy changes with the advent of new technologies over time, it is understandable that the tools we teach with will come and go. Therefore, it seems that our pedagogic concepts and teaching methodologies will come and go as well. This coming and going sometimes is more a race to keep up with technological trends than an evolution from one trend to another.

In any event, these trends affect the process of education significantly. A new media course designed specifically for production work on a tower computer will need to be altered for work on a tablet computer, which has less memory, processor speed, and screen real estate. The same course will need to be altered again, when smartphones or other devices become part of the course or when they replace the tower computer and the tablet computer in the marketplace altogether. These alterations are necessary, because the process of creating course-related work on these devices will vary from device to device.

Here, again, we see the importance of hardware and its influence on the educational process. The hardware, to a large degree, controls how educators devise learning modules for instruction. How the user or student accesses information on any particular device or hardware is determined by the physical configuration of that hardware. Teachers, therefore, are compelled to come up with new ways to plan lessons for students to access information for learning that work with the technology, both the hardware and software, they choose for the classroom.

Generally, when certain technological devices are manufactured, they have set purposes that are not necessarily in line with educational objectives. For example, a smartphone is primarily used for social interactions. Because of this, integrating devices and tools in education poses challenges. We educators, in essence, have to tweak devices and tools so that they serve our purpose, not that of their manufacturers. It is not an insurmountable challenge, as I illustrate in this book. However, it means that educators may need to adapt, even compromise, our vision for technology in education, until a time in the future when more devices and tools are made specifically for pedagogical use. That said, the way we use technology in the classroom now can and should influence how new devices and tools are created for education in the future.

Earlier I mentioned the tablet computer. At the time this book is being written, tablet computers have become a mainstay in some educational circles. Many private educational institutions—whether grammar schools, middle schools, high schools, or universities—have made tablet computers mandatory for certain studies. These devices are mobile, intuitive in terms of use, and capable of storing much data. The hard drives on these tablet computers are large enough to store an entire collection of e-books, among other data. The size, physical configuration, and computing capability of the tablet computer are important to consider when educators develop curricula that use the device. And special consideration should be given to user experience. When educators create instructional modules that rely on the swipe of a finger across a glass surface, students access information in a most facile way. This intuitive user experience may be a factor in determining how students learn with the device. Faster access to information may accelerate and improve learning, or, conversely, swiping may impede learning by allowing students to scroll by information so quickly they do not absorb it.

All of these facts were taken into consideration recently, when a professional peer of mine was tasked with developing a curriculum for a new, private school based in New York City, with plans to expand around the globe. Since the curriculum was being developed for a school that had not existed before the prevalence of new media in education, it was clear to the school's administration and faculty that technology—and the streamlined learning experience for students that can result from using technology in education—would have to be considered when devising any curriculum. After all, the school was opening in the age of new media. The specific device they considered was the tablet computer. Some of their reasons for choosing the tablet computer were that the device is lightweight, simple to use, easy to transport, and robust in terms of data storage.

The device was integrated into a particular curriculum, and all students were required to purchase it for use in their education. This would alter the process of education, which I will refer to as the *how*. In this situation, the *how* of education became device-dependent. With this mandatory device, the learning process would change for the students. Students would get the information for their education—materials, lessons, tutorials, examinations, and so forth—with the swipe of a finger across a glass screen, not via the conventional clicking of a computer mouse to access data in a folder structure on the hard drive of a laptop or tower computer. This kind of shift in the *how* in education may make machines like laptops and tower computers obsolete. Time will tell. It certainly has already made taking notes on three-hole paper and the storing of the pages in a binder obsolete, for better or worse.

Though this private school taught students ranging from pre-school up through high school, all students, no matter what their grade, were required to purchase a tablet computer. The use of the tablet computer served a worthy purpose. It enabled educators to structure the *how*, consistently across age groups. Requiring the tablet computer and developing a curriculum for it, educators put in place a school-wide teaching methodology, a foundation for Education 2.0. Teachers could employ this methodology from class to class, subject to subject. And, it was a methodology that would likely be carried over, with adaptation, from year to year. It also made for a generally consistent learning experience for students, in terms of how educational content would be presented. And in the ever-changing world of new media in education, this kind of structure is welcomed.

With a discussion of new media in education, it is easy to focus on tools, because to many people, both in and out of education, new media is all about tools. That is a straightforward assumption that we must avoid. Education is not complete until the balance between theory and practice is implemented in curriculum. In addition to the *how*, there are the important theories that support instruction—the *why* in educational practice. Theory and practice, *how* and *why*, are both important in the use of technology, hardware and software, in education. But with the integration of new technologies in education, perhaps untried and untested, *why* becomes even more important. To add further depth to the educational experience for students, we must address issues surrounding *why*.

We want our students to be engaged on many levels of thought. We do not want our students to get lost in pushing buttons when using new media. We want our students to think about what they are creating, ponder what they are experiencing, and reflect on what they are doing. We want them to fully understand that instruction with new media will support them in their educational journey, quest for knowledge, and chosen career path. An emphasis on *why* will motivate students to think, discover creative pathways in education, and find inspiration with tools. With *why*, educators provide purpose.

Certainly *why* can take many forms and focus on different things. We can use *why* to present ideas and contrast values—whether the ideas and values are simple or complex, local or global, or interesting or banal, or whether the ideas and values belong to us or someone else. It is also easy in a discussion about new media in education to focus on the tools, because the tools can be amazing and mind-bending. Technology can be magic. Swipe your finger across a magazine app opened on a tablet computer and watch the pages flip sequentially in front of your eyes. Magic! However, it is important to remember that the tools are only as good as the theories that support them. *How* follows *why*, sort of like form follows function in the design arena. This book presents a call to action for educators: Find your own ways of using new media to support a theory/practice balance. Find the magic, use it, and explain it to students wisely and completely. This will lead to a successful and purposeful integration of new media in education.

Continuing with curriculum development for the private school that requires all students to have a tablet computer, the administration of the school deemed it necessary that during the curriculum development phase, before the school actually opened for instruction, the new curriculum would be tested with a professional focus group, with participants representative of the school's incoming class of students. This was smart. This allowed the school to find and further define its purpose for using new media with its target demographic.

During the testing, it was discovered that most students who owned or used a tablet computer prior to matriculation at the school also used the tablet computer for entertainment and non-educational purposes: listening to music, accessing online videos, engaging with applications, playing games, and so forth. Few students had prior experience learning with the devices. When the tablet computers were introduced to the focus group, initially there seemed to be some confusion, resulting from a perceived conflict of purpose. When some students were asked to follow a course plan, engage with educational content, and learn new material using the device, they were befuddled. It did not make sense to them. In these students' minds, the ease of use of a tablet computer supports non-educational engagement, playing and socializing. Using the tablet computer for educational engagement, learning, was awkward. Playing and socializing came naturally to them; learning was work.

Many students in the focus group were reluctant to work at learning, even on this intuitive device, until the instructor balanced *why* with *how* in the lecture. When the teacher explained why the tools were being used, that the tablet computer was a device that could present many types of information—such as text, video, audio, and interactive graphics, which a more conventional medium such as a textbook could not—and that the information students were presented with was germane to their studies, students were more willing to engage with the tablet computers for learning. The students in the focus group realized that media literacy was as significant to learning as playing and socializing.

By supporting the use of the device with theory, the teacher provided students in the focus group with context. Context when using technology in education is crucial. Because so many devices are in our lives for so many different reasons, context defines our motivation for using them and supports the theory behind why we use them. Teachers needed to present the tablet computer in a way for students to understand that ease of use and portability are as useful for learning as playing and socializing.

When developing curriculum that is dependent on new and emerging technologies, be mindful of a cause-and-effect paradigm. When the *why* is altered, the *how* may also be altered and vice versa. Technologies change. Embrace this cycle of change in order to maintain the necessary theory/practice balance in teaching with new media. Let pedagogy evolve with advances in technology. Handheld devices and other technologies can be powerful tools in education, when students understand that with the swipe of their finger, troves of information may reveal new worlds to them, give them insights into topics they know little about, and bring them to places they have never visited. However, in the future, with the development of multimodal interfaces, students may access the troves of information with the glance of an eye or the utterance of words.

Tablet computers were not the only technologies employed in the education of students at this school. Tablet computers were the only devices required. Teachers used other new media platforms and tools—such as smartphones, apps, social networks, and laptops—when they felt their learning outcomes would be served. Because these technologies were not required, their use was not consistent and not as fully integrated into curriculum as the tablet computer. Pedagogy suffered with the use of these other tools and platforms, because the pedagogy lacked structure. The teaching/learning experience was less productive. Teachers found that pedagogy with new media is more successful when technology is required, not optional; when technology is used consistently, not intermittently; and when technology is integrated into structured lesson plans.

There is a flipside to requiring devices: not all students can afford them. All students at this private school would be required to have a tablet computer. It would seem that people who can afford tuition at a private school in New York City could afford a tablet computer. Therefore, there would be little financial difficulty imposed by the school requiring the purchase of the tablet computer. However, not all students are as fortunate as these private school students. Later in this book, I talk about a new digital divide, not one based on Internet access or availability of technology but one based on finances. When students do not have the financial means to purchase technology that is required to learn, Education 2.0 may not work. This is a challenge we must address.

There are zealots in the world of *technology in education*, who thrive on obtaining and using any new technological bell, whistle, or gizmo that may come their way. To them, education is not as solid without the gizmos. These zealots will want the gizmos forever. By contrast, there are traditionalists who view new media tools as unnecessary, even frivolous. They both have valid arguments. Sometimes new media in education works. Sometimes it does not. One of my objectives in writing this book is to show, by example, how new media technologies *should* and *should not* be integrated into pedagogy, how they *can* and *cannot* be effective. By presenting this balanced approach to new media in pedagogy, I aim to provide a framework that will shape a constructive debate, as well as put forth information that is the basis for future discourse on this topic.

Moreover, I present findings from six case studies, conducted specifically for this book, which may prompt creators and manufacturers of new media tools, devices, and technologies to fashion new tools, devices, and technologies specifically for education. That way, educators who find new media useful will not have to repurpose tools manufactured with the consumer in mind. Educators will have new media manufactured specifically with the student in mind. These kinds of new media are necessary for new education.

2

TECHNOLOGY, PURPOSE, AND MEANING

In his book, *Program or Be Programmed: Ten Commandments for a Digital Age*, Douglas Rushkoff tells us that we are part of a digital revolution in which we must "determine the value-creating capabilities of the technologies"[1] we use, not simply use them. Rushkoff puts forth that "we have embraced the new technologies of our age without really learning how they work and work on us."[2] Similarly, some instructors of new media embrace tools without really learning how they work in education and on students—how these tools are meaningfully integrated into teaching and how they affect learning. All of this should be avoided. Tools should be used with purpose and serve learning outcomes.

When teaching new media, theory and practice are best when inextricably linked and balanced, as discussed in the previous chapter of this book. Students should know *why* they are pushing buttons, not just *how* to push them. Practical proficiency with new media—that is, using a tool or technology—should have at its core the knowledge of what a tool or technology is *capable* of producing, its potential, more than just what instructions say it will produce. Guided by instructors who balance theory and practice in pedagogy, who support the innovative application of new media, and who encourage students to think out of the box, students are more apt to use technology in new ways and produce meaningful digital works that work.

---

[1] Douglas Rushkoff, *Program or Be Programmed: Ten Commandments of the Digital Age* (OR Books, New York, 2010), 13.

[2] Rushkoff, *Program or Be Programmed*, 13.

Universities and colleges with programs that focus on new media do not want to be considered trade schools. This is understandable. Therefore, striking the balance between theory and practice, *why* and *how*, in course instruction in these programs takes on even more meaning. It is a topic discussed in depth at both administrative and academic levels, and rightfully so. Both administrators and teachers have a vested interest in offering programs of study that educate more comprehensively than a trade school might. Trade schools are useful institutions, but they are primarily focused on teaching the mechanics of trades (or disciplines), whereas universities generally take a more comprehensive approach to study that comes with this theory/practice balance.

Using new media in education should begin with a foundation in academic and artistic theories. This theoretical foundation informs students' use of tools, technologies, and platforms. However, this foundation need not be rock-solid. Contrarily, it should be flexible, like the pedagogy that is used to teach new media tools should be evolutionary. Fluidity is necessary in many aspects of new media education. If the academic and artistic foundation and the pedagogy that results from it are not flexible, the process of learning new tools is arduous, routine, and boring for students. In addition, the technique taught quickly becomes obsolete. When it is obsolete, education with new media becomes invalid. This can create a kind of learning paralysis. In his book *Education Automation*, R. Buckminster Fuller summarizes what can happen when the educational process is not evolutionary or flexible.

> The new form must be spontaneously complimentary to the innate faculties and capabilities of life .... What usually happens in the educational process is that the faculties are dulled, overloaded, stuffed and paralyzed so that by the time most people are mature they have lost their innate capabilities.[3]

It is inevitable that new media tools, technologies, and platforms will continue to grow in number, scale, and complexity, and they will continue to change. When an instructor creates pedagogy that balances theory and practice and the pedagogy works, the instructor should stick with it. However, the instructor should adjust it as necessary, such as when updates of software tools are released. This flexible approach will do the following:

---

[3] Richard Buckminster Fuller, *Education Automation: Freeing the Scholar to Return to His Studies* (Southern Illinois Press/Carbondale & Simons, Inc., London, 1962), 7.

1) Create an educational experience that is current and relevant.
2) Help to prevent future paralysis of students' faculties.
3) Spur the creation of original and interesting content: "new forms."

This flexible approach also allows students in a single course, with varying skill sets and learning experiences, to benefit from a single pedagogy. As more and more high schools teach new media, students' level of proficiency with tools and technologies when they enter college varies greatly. So, professors are tasked with leveling the technological playing field, as well as teaching tool proficiency and the concepts that support that leveling.

Let me provide the example of a sequence of two courses I teach called Digital Core. This freshman-level, introductory sequence presents students with an overview of new media methodologies and practices, with an emphasis on electronic media production. The courses are required for all majors in the program. There is no placing out of the course should a student enter the program with skill or proficiency with any of the tools or technologies taught in the courses. One of the reasons for the mandatory enrollment is that the sequence is more than just an introductory course. The sequence presents an opportunity for students to meet like-minded peers in the same major: electronic media. It provides a supportive environment in which students can collaborate and create a community of peers for their college careers and beyond. And, most importantly, students share concepts and inspirations that come from their experiences prior to entering the university, as well as from each other, while learning basic production tools and techniques. This communal learning environment is fertile with new ideas. These ideas provide motivation and purpose for using software tools at this introductory level, as well as themes for projects that students will create with new media.

So, all students in the major take the courses, but inevitably skill sets and capabilities vary from student to student, not only because of prior education with new media at the high school level, but also because some students simply learn the tools faster than others. This is one instance where striking the balance between *why* and *how* in instruction is definitely necessary. The students who are or quickly become proficient pushing the buttons—let us call them the technicians—may not inform the use of the tools with interesting purpose. They may not have the proclivity to think out of the box. The others—let us call them the creative thinkers—may have grand vision, but they may not know how to manifest it because the software is new to them, or they have little experience pushing the buttons, or they are challenged learning the tools. We want students to be both technicians and creative thinkers.

In this one sequence of courses, there are two groups of students ready to begin their study of electronic media with different skill sets, diverse experiences, and disparate production capabilities. Thus I needed to find a way to begin instruction that was fair to all students. In order to help resolve the imbalance, I created a series of collaborative exercises in which I teamed up students with various skill levels. My strategy was that students would share ideas and concepts with one another. Therefore, what they were unable to learn from me, in terms of what to do with the tools or how the tools work, they would learn from each other. In my experience teaching new media, I have found that teamwork and community building are both essential, when there are inequities of skill and experience among students. In the case of Digital Core, teams worked.

On some level, teaching Digital Core is like teaching many courses in one. There are a number of instructional objectives to meet: software instruction, theoretical explanations, community building, and the presentation of case studies targeted to students who know, students who do not know, students who can, students who cannot, students who are motivated to learn, and students who need motivation to learn. This is a real mix—a mix that is common in beginning-level classes in new media disciplines across institutions of higher learning, public, and private. It is also indicative of the real world. I have professional experience outside of education as a director of new media for Fortune 500 companies in New York City and for a global foundation founded by a former president of the United States. In my professional experiences, some people knew the stuff, and some people worked hard to convince other people they knew the stuff. There was a real mix.

When instructors establish *purpose* first, it is easier for them to teach the tools, for students to learn the tools, and for students to understand why they are using the tools. *Purpose* becomes the objective for using the tools. When working with new media technologies, students need to be reminded throughout the learning process that *why* is equally important to *how*. When instructors provide clear instructions and support their instructions with tried and true theories, academic and artistic, students are likely to produce digital projects that matter—however *matter* is defined. *Why* is not more important than *how*, especially at the beginner level. Balancing the two is the real goal.

Anyone can learn the tools; it is what you do with the tools that matters. Therefore, *why* does edge out *how*.

In order to provide students with solid purpose when teaching new media tools, I incorporate service learning and community outreach objectives into my assignments, when possible. My goal is to involve students in partnerships that I have created with not-for-profit groups, such as charities and arts organizations. This way, I establish *why* inside the classroom by engaging students with community outside the classroom. I guide students in the production of projects that give back, which make a difference, which matter. Projects in Digital Core include websites, video public service announcements, podcasts, and simple, educational games. As this is an introductory sequence of courses, students' projects are basic in terms of production quality, but the experience of creating the work is far from basic. Working to serve the missions of not-for-profit organizations, students produce projects that help disenfranchised people, create awareness for artists and art organizations, communicate the messages of non-governmental organizations (NGOs), and promote social causes.

To create new media projects in Digital Core that will help society, students need to learn to know what they are doing—how to design websites, edit short videos, produce podcasts, and even design basic interactive games. Throughout the process of teaching new media, I remind students that the work they are creating with machines can make a difference to human beings. Once the tools have been studied, the lessons learned, and the works produced, students' fully realized projects take on *meaning*, derived from the established purpose. Websites are launched, and videos and podcasts are uploaded to the Internet. These projects help share the mission and work of the not-for-profit organizations with people around the world. Purpose has advanced to meaning. Meaning is significant. Meaning has staying power.

If educators want to make sure that the universities and colleges where they teach new media are not trade schools, the tools they choose for instruction must be used with purpose. Again, *why* edges out *how*. In the case of Digital Core, I support *why* with the presentation of the theories of digital storytelling, human computer interaction, aural communication, and interface design. When these theories are successfully integrated with practical know-how, the resulting video narrative, podcast, website, or game has meaning. The student, who understands this integration and produces meaning in the classroom, will more likely be successful in any field that has new media, and most all professions do these days. However, I do not want to limit the definition of *meaning* to only work for the not-for-profit sector. I believe that commercial work can also be meaningful, just in a different context. The purposes for selling products, persuading voters, marketing concepts, celebrating celebrities, and publicizing events are not worse. They are not better. They are just different.

We educators have an obligation to expose our students to an all-purpose way of using new media: educational, cultural, artistic, *and* commercial. For me, this comes later in the high-level courses I teach, after students have established a sense of community with their peers, gained confidence from working collaboratively as well as independently, acquired a certain level of proficiency with new media, and realized that what is produced with the tools is more important than the tools themselves—usually.

3

TOOL LITERACY

We are at a point in the digital revolution where technology is integrated into many aspects of our lives, especially education. To think that using software, hardware, online tools, and digital devices in education is unnecessary, experimental, or frivolous is shortsighted, limiting in terms of creative and academic possibility, and, more straightforward, old school.

Tools, both hardware and software, *should* be used in education. They facilitate a new kind of learning that is right for our digital age. It is education that embraces the purposeful use of technology in pedagogy. It is new education. Software tools like the Adobe Creative Suite, tablet computers like Apple's iPad and Sony's Tablet, and e-readers like Kindle and Nook provide new and effective means for teaching students. Because of their widespread use in education, it is obvious that these tools and technologies are here to stay, in some form or another. Therefore, we just need to carefully examine how they work or, better, how they *should* work in education.

I advocate for *tool literacy* in higher education. Tool literacy guarantees that students graduating from institutions of higher learning will possess a broad skill set for using new media tools, a skill set that is required in many professions changed by the digital revolution. Tool literacy is not limited to use of the tools. It includes understanding the theories that inform the meaningful use of these tools. Again, this is the theory/practice balance mentioned earlier. Tool literacy begins with carefully formulated tutorials that are created by instructors. These tutorials should teach students tools, while focusing on specific course objectives and meeting learning outcomes. Instruction should emphasize *why* as much as *how*, as already established. Over time, students acquire a level of proficiency with the tools, as well as a deeper understanding of what they can accomplish with the tools. This combination of proficiency and understanding are the two main components of tool literacy. There is a third: curiosity. We need to encourage students to find new ways of using tools, to create original digital projects, interactive experiences, and multimedia narratives that heretofore have not existed. With proficiency, understanding, and curiosity, students are able to successfully use tools and technologies in a variety of areas.

More and more, new media tools and technologies are used across disciplines. The tools used and methodologies employed to manipulate digital images, design user interfaces, edit video and audio, animate content, and produce interactive experiences are as common in electronic media as they are in health care. Certainly, there was a time when graphic designers, video editors, animators, interaction designers, and multimedia artists were the primary users of these tools. That time has passed.

New media are pervasive today. Journalists, literary experts, engineers, architects, scientists, and physicians also use them, and the list of professionals is growing. The current inter-disciplinary demands on study at universities, as well as an inter-disciplinary focus in professional environments and the need for proficiency with new media tools in both education and the professional sector, are all reasons to require formal instruction in industry-standard tools in the classroom and to make tool literacy a priority.

By making tool literacy a key pedagogical objective and an educational priority, educators rightfully establish a need to teach tools in higher education. In a production-oriented discipline like electronic media, it is sensible and necessary to integrate software tools into classroom instruction, because the use of software and hardware is commonplace in content creation in the professional world of electronic media, as well as in related areas. Next I present additional examples of how I have integrated tools in my own courses—a process of defining purpose, balancing theory with practice, and guiding students to make meaning. These examples also present some of the challenges I have faced in the ever-changing new media space and what I did to meet those challenges.

In a new media course focused on interaction design and involving the design and production of user experiences in screen-based environments, I use Adobe Flash. Flash is an industry-standard tool for creating interactive and animated content online. Interactivity in Flash projects is authored with the scripting language ActionScript. During a recent upgrade of Flash from version 2.0 to 3.0, Adobe significantly changed ActionScript, in terms of its algorithmic logic and scripting syntax. As a result, the learning curve became steeper, especially for undergraduate students. The tool became a more challenging one to learn and to teach. More time was required in the classroom to teach the students *how* to use the tool, resulting in less time spent on covering theory and addressing *why* they were using the tool. Therefore, the software upgrade caused a rethinking and reworking of the course modules, as well as a tweaking of the course's learning outcomes. This kind of revamping of a course is necessary and a real by-product of technological advancement.

It is important to point out that instructors may not need a software upgrade to teach students how to use tools with purpose. *Why* is not dependent on *how*. Theories of interaction design are the same with ActionScript 2.0 or 3.0. In short, interaction design has its theoretical foundation in human computer interaction, or HCI, which, as its name suggests, defines how and why we engage with computers and work with the intelligence that computers provide us, in order to meet objectives and accomplish goals in our daily, human lives. If students can become proficient with Flash and understand the important applications of the tool more easily with version 2.0 than with version 3.0, 2.0 may be better for teaching interaction design.

In another of my new media courses, I teach students visual communication in web design. In one particular module, students create an interface design composite of a website for a fictitious horror film. Given the genre—horror—the composite should be stylized—horrific. Working in Adobe Photoshop, students were instructed to visually manipulate imagery using some of Photoshop's many artistic filters. Horror is broad, so almost anything scary, grotesque, weird, or just out of the ordinary might work. But this horror film, which had yet to be released (per my instruction), was based on a novel. Students were required to read the novel, interpret its depiction of horror, which they were told was paramount to the project's visual communication, and then redo their composites to work with the overall visual theme. The novel's depiction of horror became their design purpose.

The focus of yet another new media course that I teach is the creation of visual effects to support narrative in short digital films. The tool used in this course is Adobe After Effects. For a desktop tool, After Effects gives video artists a wide range of effects and filters that can be applied to digital video footage to enhance narrative. It is inevitable that narrative structure will evolve, as new media technologies provide unique platforms for storytelling.

The small screen of a handheld device, for example, does not and cannot deliver the visual impact of a 40-foot by 23-foot theater screen. Therefore, as more original content is created for the handheld screen—instead of content being repurposed and resized from the big screen to fit the handheld screen—visual effects become necessary to convey emotion, provide visual continuity, and ultimately keep the audience involved in the story.

In my visual effects course, students are required to create short films in which an original narrative, which they devise, is supported by visual effects that they create with the tool After Effects. In various exercises, students illustrate how visual effects help to develop plot and engage viewers in their story. Then, in a final presentation of their video works, students defend their choices of visual effects and explain, one by one, how each effect purposefully serves the narrative. The result is that, because of a software tool, students acquire a level of visual literacy and an understanding of the importance of visual imagery in digital storytelling.

The author Shilo T. McClean is an expert on visual effects theory. In her book *Digital Storytelling: The Narrative Power of Visual Effects*, McClean asserts that digital effects are fast becoming an important component of story development in contemporary film and video art. She urges young filmmakers and new media artists not to "underestimate the scope and power of digital effects."[4] I share this quotation with my students, while emphasizing that a powerful story can be enhanced by powerful visuals. Enhancing narrative is yet another way students can create meaning with new media and software tools. Students learn that visual effects can be integral to presenting narrative, and they learn a new way to communicate their stories.

The process described in the previous example is another illustration of the importance of balancing theory and practice in instruction. The computer hardware students edit on, the software they use to edit, the technology they use to add visual effects, and the digital compression of the video for export are all practical components, or tools. However, in order for students' stories to come to life meaningfully, students must have a grasp of the fundamentals of storytelling, such as how to create a narrative arc and develop characters, as well as an understanding of the impact of visual effects on the presentation of a story, as well as the audience's interpretation of that story. These are the *theoretical* underpinnings of the exercise. By combining *how* with *why*,

[4] Shilo T. McClean, *Digital Storytelling: The Narrative Power of Visual Effects* (The MIT Press, Cambridge, MA, 2007), 34.

throughout the process of creating stories—shooting video footage, editing that footage, and exporting the finished video—students produce a complete project. Purpose evolves into meaning. And, more importantly, students learn the "value-creating capabilities of the technologies."[5]

New media tools create a foundation for new education and a necessary way to teach a wide variety of disciplines, from electronic media, to engineering, to English, to journalism, to medicine. Yes, almost every discipline that is taught in higher education, whether theoretical or practical at its core, can benefit from instruction with new media tools. I would even go further and say that all disciplines taught in higher education should require some instruction with new media tools. The integration of tools in pedagogy across the board and across curricula will have students producing work in the classroom that is relevant to the technology-focused world outside of the classroom. Why? Because tools are a part of the real-life application of these disciplines.

It easy to understand how tools are part of professions related to electronic media, engineering, and other disciplines rooted in technology, but how are tools significant to disciplines considered more traditionally academic—like literature, medicine, and natural sciences? And, how are new applications of new media integrated into the performing and visual arts? To answer these questions, let us begin with the example of an English professor who teaches British literature and who wants to introduce his students to a timeline of British authors from Victorian England. What are the options for creating such a detailed timeline?

I think both proponents and opponents of technology in education would agree that drawing the timeline on paper and photocopying it for the class of students would not be an option. The days of hand-drawn diagrams photocopied for distribution are far behind us. The timeline could be created in a word processing program, like Microsoft Word, and distributed as a list of words with simple line diagrams. But in its presentation, the timeline would be two-dimensional and flat, and thus the content, though perhaps spot on and substantive, would appear lackluster. I do not want to suggest that appearance should have anything to do with substance. However, in our visual-centric, media-focused culture, it does.

And there certainly would be no interacting with a timeline that was two-dimensional and flat. If the work were interactive, the content would be even

---

[5] Rushkoff, *Program or Be Programmed*, 13.

more compelling. Therefore, a better solution would be to create an interactive timeline in a tool like Adobe Flash. Such a timeline could show where the British authors came from, the years they lived, the works they published, as well as video and audio clips documenting passages from the authors' works—as dramatization or narration. Students' learning experiences would be made richer by this kind of interactive experience.

More technology could make the interactive experience even richer. For example, the user could click on hyperlinks that would access online information pertaining to the geography of Britain, historical events of the time, works of contemporaries, and so forth. Flash is a robust media production tool that can produce a timeline, which is deep with layers of content and media assets. With proper back-end coding, the tool can integrate graphical user interfaces with databases for connections to useful repositories of data. Tools like Flash provide a learning experience that engages students and enhances students' education, in a way that a drawing or a word processing document simply cannot.

Aside from the presentation of content, there are two questions pertinent to the previous example that should be answered:

1) If an English professor wants to engage students by using a media-rich, interactive timeline created with Flash, why would he or she need to learn Flash?
2) If an English major plans to write novels or go directly on to graduate school for more academic study, for example, to pursue a doctorate degree in English, why would he or she need to learn Flash?

It would be easy to say that these individuals do not need to learn Flash. Proponents of across-the-curriculum integration of new media might make the argument that if either the English professor or the English major needed a Flash piece for his or her work, he or she could partner with someone who knows Flash, like an electronic media student or a local new media professional. It is not that simple any more.

The integration of digitally produced and technology-driven content in most professions, including ones that have historically been academic in nature, requires anyone who wants to land an entry-level position, pursue further study, or attain a modicum of success in his or her career of choice to know how to work with industry-standard tools and technologies, at least with minimal proficiency. This is due to the evolving interdisciplinary nature of many professions, a global reliance on technology, as well as the economic realities of certain job markets. For example, it is not unusual today for a person with a doctorate in English, who takes the position of digital associate editor with an international publishing company, to write blog posts, produce videos, assemble photo galleries, and curate online content. In order to be competitive, students with university degrees, both undergraduate and graduate, will likely need a command of new media tools upon entering their chosen professions.

Following are several examples of students of mine, who have advanced in their careers because of their knowledge of tools and technologies and their new media expertise. It is important to point out that not all of these students majored in a new media-related discipline. However, they all saw the need and had the foresight to learn valuable *digital* skills before graduating.

1) An engineering student took a digital imaging course I taught. During his last term at the university, he interviewed for a position at a reputable engineering firm. The firm narrowed the search down to two candidates. He was one of them. My student got the job because he knew how to use Photoshop.
2) A student of mine was doing post-graduate research at a premier cancer hospital in New York City. His boss, the lead medical researcher and a well-known research physician, required that his students learn how to use the Adobe Creative Suite, so that they could document their research with poignant visuals and animations. My student, who was already familiar with the Adobe Creative Suite, had a big advantage here, and he was able to produce compelling visual documentation of his research without added cost to the hospital.
3) An electronic media graduate who studied under me went to Los Angeles wanting to work in the media industry. Frustrated by the competition and not having the contacts that could assist her in landing a job with a major studio, she applied for a job at a boutique production company. She was offered a position working alongside top management, because of her knowledge of digital media tools and her electronic media production skills. She became a go-to person

> for projects that required someone with new media proficiency. This put her well on the way to a career in media in Hollywood.

In addition to specific examples of individuals relying on tools to procure employment, entire industries are being reinvented, becoming more reliant on digital tools. Industries that historically have not used new media or have been reluctant to integrate new media into professional practices are more and more looking for qualified persons who know how to work with new media technologies, tools, and platforms—sometimes even expecting entry-level employees to have proficiency with industry-standard tools.

One such profession is journalism. As of the writing of this book, journalism is trying very hard to reinvent itself. Its business model of printing news and selling it on paper is obsolete and no longer financially viable. The traditional means of distributing the news has been eclipsed by the onslaught of digital media providing news in a 24/7 cycle on the World Wide Web and via customized applications. Sales of paper copies continue to dwindle. Most newspapers and news magazines have an online presence and are relying on other media, such as mobile and handheld devices, as way to reach a contemporary audience who want their news in a digital format. Journalism needs graduates with digital know-how.

Web 2.0 technologies have given birth to a new kind of news, one that is media-rich, interactive, and democratic. Because there is a preponderance of devices on which to access the news, and there is no single model for a successful news website, mobile site, or application, producers of news are creating content that will engage users in a way that keeps them coming back. Repeated user visits are good for advertising revenues. For example, *The New York Times* is represented online with a complex website that presents news on a whole new level—with multimedia, video, interactive graphics, audio clips, slide shows, databases of information, and interactive blogs. Stories are not just told with words. Stories are told with a combination of text, media components, and interactive features. News is not just read. News is experienced.

Bringing this back to higher education, the journalist that instructors would educate to work in this technology-driven environment is the journalist who has an equal command of writing and new media. Let us call this professional a *new journalist*. The instructors who prepare the new journalist for a successful career must understand the value of tools in education and industry, as well as the necessity of acquiring proficiency with the tools.

In some situations, namely smaller markets, the new journalist may be required to design and produce media features and interactive components in order to communicate the news. Or, the new journalist may work with a team who will design and produce the digital content. In either case, new media technologies are now an integral part of telling a news story, and in order to successfully tell a story in the online world of journalism, the new journalist must know tools. The new journalist is representative of many *new professions* created by the ubiquity of new media.

The medical field is also embracing new media in a way that goes well beyond the use of technology for scientific research. I collaborate with medical researchers, who want their work presented in various new media formats—interactive kiosks, online platforms, videos, audio podcasts, and so forth—in order to get their research out to the scientific community and the public in a quick, efficient, and compelling way. Medical professionals realize that new media can expose a much larger audience to their work than a traditional print journal, magazine, or newspaper, and that a media-rich website is more compelling, in terms of user engagement with content.

For example, an interactive component or multimedia feature that immerses the user in new findings about the frontal lobe of the human brain will be more captivating than a static, two-dimensional diagram, because the user could explore deep inside the brain to experience the science. Learning becomes experiential.

New media technologies are a win-win in science. Certainly new media can engage scientists and learned people with facts and data in a fresh way, but they also have the power to spur interest in non-scientists. In the long term, new media can be more cost effective than other media, because digital content is easier and cheaper to update than printed materials or broadcast content, in the case of television. In addition, the frameworks for this digital content—websites, kiosks, and so forth—is scalable. This is especially helpful for laboratories whose research results vary and change over time, requiring periodic updates, and whose funding is limited. Also, using new media to create or expand awareness of their research can help research scientists obtain more grant funding and interest accomplished scientists in joining their teams. The proper preparation and production of scientific research for display on the World Wide Web and in other digital arenas require an expertise that comes from tool literacy.

Technology and software have also become an integral part of producing the performing arts. A visit to any major Broadway theater or international performing arts venue will reveal how video projections have been integrated into theater, opera, and dance with visually stunning results. Recently, I taught a sequence of courses that had students working with me to create video projections for a main stage opera. Students were involved in a complex process that began by attending concept meetings with the entire creative team (director, designers, and production personnel), continued with the shooting, editing, and producing of large-scale, digital moving imagery, and concluded months later with the presentation of high-end, artistic video projections onstage.

Many new media software and hardware tools were involved in this collaborative and creative venture: Adobe Photoshop and Illustrator for the creation of storyboards; high definition cameras for the shooting of original video footage; Apple's Final Cut Pro for non-linear video editing; Adobe After Effects for the creation of visual effect; and Dataton's Watchout, a media delivery system used to integrate the imagery into the stage setting and programmed to present the video projections at specific cued moments throughout the opera. This experience showed students how new media can be presented in live performance, outside of a screen-based environment. And it exposed students to another potential career in new media that requires tool literacy: projection designer and producer.

The importance of technology in education, in terms of creating methodologies that will prepare students for careers in today's digital environment, cannot be underestimated. I continue to make the case that understanding and knowing tools is important, not necessarily because I advocate pushing buttons, but because today, proficiency with tools is required to present new ideas in academic, artistic, and professional worlds. It just is. New ideas, like those of my students, will shape education, arts, and industry in the future. Whether a student is inspired by an industry executive demonstrating the newest tablet computer in a web video, an English professor illustrating the interconnectedness of Jane Austen and British society of the late 18th century in an interactive Flash timeline, or an opera performance impressing an audience with high-tech, moving imagery, technology serves an apparent purpose. It brings us closer to and deeper inside our technology-focused world.

It is inevitable that the tools and technologies available to us will continue to morph and change in order to meet the demands of users and marketplaces around the globe, in and out of education. It is up to us educators to remain flexible in our decisions regarding technology and our implementation of tools and to adapt our teaching practices as tools change. That way, we may forge ahead with ingenuity and create educational experiences for our students that are impactful, with applications of new media that make meaning.

In 1990, psychologist Jerome Bruner gave a lecture at Harvard University entitled *Acts of Meaning*, in which he argued that meaning should play a role in the human learning experience. In his lecture, Bruner presented research findings that supported the notion of a cognitive revolution, with a fixation on mind as an information processor. Bruner, a somewhat disillusioned scholar at that time, stated that this fixation had led psychology away from the deeper objective of understanding the mind as a creator of meanings:

> [The aim of the research effort] was to discover and to describe formally the meanings that human beings created out of their encounters with the world, and then to propose hypotheses about what meaning-making processes were implicated. It focused on the symbolic activities that human beings employed in constructing and making sense not only of the world, but of themselves.[6]

If we apply Bruner's conclusions to our integration of new media in education today, we can learn from his research. Let us strive to employ *meaning-making processes* in instruction. Let us teach tools, but let us do it with an edge toward *why*. So whether the overriding goal of education is to prepare students for a profession, or to show them how they can make a difference, or both, let us teach students to create new media works that reflect their human experiences, use themes that are relevant to contemporary times, enrich our universal repository of knowledge, and help us make sense of our new media world and ourselves using new media. Then, tool literacy is part of a meaning-making process.

I believe all of us, students and faculty alike, can make meaning from our vision. With tool literacy, we are better equipped to do so in these new media times.

---

[6] Jerome Bruner, *Acts of Meaning: Four Lectures on Mind and Culture (Jerusalem-Harvard Lectures)*, (Harvard University Press, Cambridge, MA, 1990), 2.

4

INTERACTIVE LEARNING

The use of technology in education creates an interactive exchange of information—from teachers to students, and from students to teachers—that reinvents the way we educators teach. The conventional *stand and present* model of teaching gives way to a *stand, present, and interact* one. In the former model, students receive information, digest it, and react to it. This is a process of hearing lecture material on a particular topic, taking notes as study material, and offering verbal comment in class as part of a discussion or critique. This is conventional class participation. In the latter model, students receive information, digest it, react to it, and *interact* with it. This process engages students significantly more. Students still may react verbally, such as in a standard question-and-answer scenario. However, with this new model, students engage with electronic devices as well. Class participation happens via mobile devices, tablet computers, smartphones, and the like—in addition to and sometimes instead of verbal offerings. This is the fundamental principle of *interactive learning.*

To further differentiate the two approaches, let us define verbal involvement as *reaction*, and participation via devices as *interaction.* In a single class, reaction and interaction may occur simultaneously or separately, but in order for students to have a successful instructional experience in which they will learn new material using new media devices, both reaction and interaction should be present. It seems quite basic, but, in short, students should both talk and use devices. Interactive learning does not and should not preclude verbal involvement of students. Verbal involvement in the classroom is necessary and can be most effective when combined with participation via technology.

When some material will be spoken and some material will be experiential, instructors must devise a way of using both approaches cohesively. The objective should be to integrate the two approaches and create a singular classroom environment in which instructors and students work seamlessly toward achieving learning outcomes with and without technology. To do this, an instructor needs to assess how course material is most effectively presented: verbal or interactive. Some material is better presented one way or the other. And in certain instances, when material is complicated and challenging for students to comprehend, it is best presented both ways—where verbal instructions support interactive or vice versa.

From this book's previous chapter, "Tool Literacy," let us return to the example of new, scientific research presented in an interactive format versus a static format. Here it was clear that an interactive component documenting the scientist's research would engage a user, or student, more so than a two-dimensional diagram, because the interactive component would create an experience in which the student could explore deep inside the brain, virtually. The student would have a rich learning experience, and the sense that he or she was immersed in interesting and complex content would facilitate the learning of that content. Interactive learning was the better approach here.

There are numerous factors to consider when deciding to use interactive learning:

1) Devices and technologies that are available for use in instruction.
2) Students' aptitude for using devices.
3) Students' inclination to engage with content presented on new media devices.
4) The format of the course material. Does the course material exist in a digital format that can be presented on a device, or does the material need to be converted to a digital format? There could be significant cost and time associated with converting materials into interactive or multimedia pieces.

When using interactive learning, the pedagogic objective is to engage students with instructional content that is *both* informative and experiential. If the interactive component that immerses a student in scientific research of the human brain merely takes the student on an interactive tour of the brain and does not present new and compelling information about the brain, the experience may be entertaining, but it would not necessarily be educational. It stands that some interactive learning is most impactful when supported by other teaching means or platforms: lectures, printed tutorials, textbooks, and so forth. And with the ongoing development of new media platforms for teaching, there is likely a single, sophisticated, robust interactive component in the marketplace that can *both* entertain and inform—effectively and successfully. The challenge—in the Wild West of new media where things come, go, and change constantly—may be to find that component.

Integrating teaching platforms, course modules, and class exercises to accommodate both reaction and interaction in the classroom will result in optimal overall learning by students. When using various teaching platforms, it is important to thematically connect content across platforms, so that there is a uniform teaching/learning experience delivered by the instructor and received by students. Students want to know that the material they are learning from various means is leading to one, focused end. The emphasis on interaction is likely the hook that will get students interested in new content initially, but it is the consistency in theme of subject material across teaching platforms that will keep students engaged between tutorials and make the new information they learn last over time.

Here is an example from my teaching. In an upper level course of mine focused on user experience in a screen-based environment, I alternate lectures about the application of activity theory—in which I explain why human beings choose to engage with something (theory) and how a screen-based user experience is crafted to be interactive and satisfying (practice)—with hands-on tutorials showing how to create interactive components for websites, using scripting languages like JavaScript and HTML 5.0. After students create the interactive, screen-based works, they load them on devices to see how the components function in a practical environment.

This course is part lecture, part presentation, part laboratory, and part experiment with new technology. All the parts are focused in one direction, and, therefore, create a singular learning experience. This multitiered approach involves students with the course material in both theory and practice, and it does so with a consistent instructional objective, or theme, from beginning to end: teaching user experience in a screen-based environment. I have found this approach successful when teaching advanced topics.

Interactive learning depends on the use of new media devices. However, there are advantages and disadvantages to integrating devices into the classroom experience, especially when interactive learning is compared to more conventional types of learning. One main issue is the digital nature of the data that these devices make and store. Because data sent via devices, which are connected to a network of computer servers and communication satellites, are digital, the data can be stored, archived, and when needed, accessed in the future. This is advantageous for students. It allows them to review lecture material, prepare for an examination, and revisit topics of interest. For example, with Blackboard I can upload and store documents such as assignment notes and study guides, which students can use as reference material to study for examinations and/or produce class projects.

But the long-term storage of data can also be disadvantageous for students. Data that is incorrect or even embarrassing—in the case of the student who offers erroneous or inappropriate information—can also be stored and potentially last forever, resurfacing at inopportune times. For example, students may spontaneously post irreverent comments on web community boards, online forums, or social networks. Once posted, it may be difficult, and perhaps impossible, to retract or delete those comments. The unwanted comments could resurface when a student applies for an educational scholarship, graduate school, or even a professional job.

The pros and cons of storing digital data should be seriously considered before educators implement an interactive teaching model. However, the potential cons aside technology—with its ability to store data and provide students with a means of participating in the learning process again and again, in a quick and facile way, and with a device that is handheld, or mobile, or both—are good for education. Moreover, the impact of interactive learning on education, because of technology, is consequential. As already established, because interactive learning is experiential, class participation goes far beyond a student raising a hand to answer a question or contribute a comment. Therefore, once we educators introduce technological devices into the learning process, we must accept the fact that we will likely need to teach students—perhaps even recondition them—to use devices in education in a new way: to learn, not to socialize or play. This requires breaking old habits, altering behaviors, and presenting a new purpose to students: devices for educational use only. This can be challenging.

Let us look at another example of technology in the classroom: technology that is not integrated into the pedagogy but is still present in the classroom. This is the case of a student sitting in class with a personal smartphone. If the smartphone is turned on, the class will be disrupted when the phone beeps and the student looks down to read a text message, assuming it is a social text and has nothing to do with instruction.

> *Beep beep beep.* On the smartphone the message reads: omg! did u c him? he is not my bff, u r. l8r.

The instructor asks the student to shut the phone off and put it away. Then, the instructor admonishes the student for interrupting the class. Most instructors do not view texting as a meaningful educational tool, and the generally accepted purpose behind texting is to communicate casually about everyday topics, not serious or scholastic ones. Therefore, it follows that unless instructors could be sure that students would learn with their smartphones, not socialize with them, those instructors would be reluctant to use smartphones pedagogically.

But what if there were a way to make the text message relevant to instruction, a way to make the text integral to the classroom experience, so that the *beep beep beep* sounded a level of participation, a component of an interactive learning? There is. The instructor could use texting to deliver content, engage students in the educational process, and even aggregate and archive information pertinent to course study. The instructor would simply have to make the choice to do this, set parameters for the use of smartphones in the classroom, and teach with the devices. Then, the arrival of new information on the student's device could be for learning, not socializing.

> *Beep beep beep.* On the smartphone the message from the instructor reads: In his book *Blink: The Power of Thinking Without Thinking*, Malcolm Gladwell encourages us to "think without thinking"; what does he mean by this?

Now, it is different. With this single question appearing on the screen of the student's smartphone, the text message is a deliberate part of the class and a necessary component of pedagogy. A new kind of educational process is underway, an interactive one—one that uses a new media device to engage students with course content and extend learning beyond the confines a brick-and-mortar classroom. And because smartphones are becoming more and more pervasive in our Western culture, it might be easy to require students to have a smartphone for class use. However, making any hardware or software mandatory for education can still be an issue for certain demographics of students. This is apparent in the discussion of the new digital divide based on finances, later in this book.

In one class of mine, I sent the exact text message mentioned previously to my students at the end of a lecture I gave on the philosophies of contemporary author Malcolm Gladwell and how his ideas apply to the study of new media. In his book *Blink: The Power of Thinking Without Thinking*, Gladwell discusses topics such as *thin slicing* and *adaptive unconscious*, to illustrate how generally we should trust our first impressions and go with our gut instincts.

> Thin-slicing ... is a central part of what it means to be human. We thin-slice whenever we ... have to make sense of something quickly or encounter a novel situation. We thin-slice because we have to.[7]

In my lecture, I talk about how Gladwell's theories apply to learning new media technologies, because in essence new media—the discipline as well as the technologies—are unpredictable and forever changing. In new media education, a person often has to rely on gut instincts when learning new new media technologies, because there may be little or no information about tools that precedes classroom instruction and little or no documentation that instructs in the formal use of the tools. It is common to find ourselves in situations where "we have to make sense of something quickly."

Sending the previous text message was relevant to my classroom instruction. It was the thesis for a research paper, "New Media and the Power of Thinking Without Thinking," which students were required to write. This thesis came directly from the subtitle of Gladwell's *Blink*. With device in hand and thesis displayed, students were ready to begin their research. There would be no need to download a file from an e-learning system or get a paper handout, because I appended my text message with the URL of a website, where online students could view detailed specifications of the assignment. I created a process for answering students' questions. If students had a question before leaving class, they were encouraged to ask it. If students had a question while they were conducting their research—whether in a library, online, or at a remote location—they were encouraged to text it. Yes, text it.

---

[7] Malcolm Gladwell, *Blink: The Power of Thinking Without Thinking* (Little, Brown and Company, New York, 2005), 43–44.

The integration of Gladwell's philosophies into my Digital Core course is discussed in more detail later in this book in "Case Study No. 2: Interacting with Literature on Facebook," where *Blink*, and its relevance to new media, took on even more significance.

It is important to note that in the particular exercise described earlier, specific text messaging rules were in place and would be enforced, in order that this exercise would have structure and integrity from a pedagogic standpoint. When establishing these rules, the issue of immediacy was the first one I addressed. Whenever electronic devices are introduced into instruction, there is an expectation of immediacy. Students may text a question and expect that the answer will come within minutes, if not seconds. There are advantages and disadvantages to this *expectation of immediacy*. One advantage is that students can progress through their research at a fast pace, not hindered by the necessity of physically meeting with an instructor to ask their questions. One disadvantage is the assumption that instructors are available to answer students' questions around the clock or whenever students are conducting research and choose to send a text message. Instructors are not this available, and they should not be.

There were other means of communication to consider as alternatives to text messaging. Telephone and e-mail might be two. However, for better or worse, telephone communication has fallen out of favor. E-mail takes time. E-mail messages are sent and received in a cycle that requires more steps—open, respond, and send—and e-mail is a method of communication made slow by web server processes and traffic on the Internet. Contrarily, my smartphone will alert me to new text messages instantaneously. If the objective is to provide for quick communication and succinct information exchange—one that creates a stream of questions and answers that can be read, reacted to, interacted with, and archived quickly—text messaging is better.

Texting also appears in a physical screen environment with fixed dimensions. This restrictive screen real estate compels the person who is texting to limit the length of a message. Since there would be a large number of text messages coming my way, from a class of 30 students, brevity of message would matter. It would help me keep things organized on my end, and it would allow me to engage in back-and-forth communication with students more easily and quickly. Beyond the classroom, this limited screen real estate has also formed how text messaging is used in everyday life: Short bits of information are presented in an invented phraseology of words, acronyms, and abbreviations.

In order to make this device-enabled research exercise a productive one, I put parameters in place and enforced a set of rules. These rules established the times when students could text and when I would respond, as well as a maximum length for individual text messages. These rules would add decorum and structure to the process and would compel students to think before texting their inquiries. The goal here was to make students' texts more relevant to study and more poignant as communication messages. Having these parameters in place also helped alter students' previously learned texting behavior by reinforcing that texting in this exercise was for learning, not socializing or playing. Students were receptive to this *texting decorum*, and they adhered to the rules most of the time.

In this exercise, students' use of texting during the research process created an archive of questions they asked and I answered. This question-and-answer stream, stored on their smartphones, provided students with a list of information for future reference—an electronic repository of information pertinent to their research that was easily available to them. Providing students with the opportunity to ask questions verbally during class and via text message after class, as well as having students carry the question-and-answer stream on their handheld device, illustrates the meshing of the *reaction* and the *interaction* components of this model of interactive learning.

In addition, using electronic devices to support research in and out of class and on a flexible schedule—allowing for optimal use of time, energy, and resources—created a foundation for device-enabled research that I will further develop in the future, when new and more sophisticated devices are manufactured and become available. There is more information on the use of smartphones and text messaging for research later in this book in "Case Study No. 5: Conducting Research with Text Messaging," where text messaging was the foundation of a more involved research process.

When using smartphones or any device in education, another matter needs to be addressed: ownership. Who owns the devices? And are the devices going to be used for purposes other than research? It is likely that students' smartphones will serve several purposes, in and out of the classroom. The phones will be used for personal, educational, and perhaps professional reasons. Therefore, when using smartphones for research, all of the text messages stored on the devices will need to be organized, so that the research stream of messages is segregated from other miscellaneous text messages on that phone. Thus, the integrity of the research will not be compromised, and the data itself will remain intact. This will keep the information in the research stream on point and allow students to locate and reference information relevant to their research easily. This need for organization of data extends to instructors, who are also likely to use smartphones for more than one purpose and may need to use their own phones in teaching exercises. One way to guarantee a separation of data would be to require that students and instructors have separate, dedicated devices for class use and personal use. This brings up the issues of cost and affordability, which is discussed in more depth later in the chapter "The Cost of Technology."

Interactive learning, with its integration of handheld devices in instruction, is one way technology is having an impact on education. The debate over whose responsibility it is to provide devices, handheld or otherwise—as well as matters pertaining to ownership of devices, access to content on devices, and the safe keeping of that content—is important to the development of new education. These topics are relevant to how we educators effectively incorporate new media devices, the content they produce, and the data we store on them into teaching. The discussion of these topics, some of which are discussed in more detail in this book, should continue. Still, all things considered interactive learning, with new media devices at its foundation, is transformational.

5

PARTICIPATORY PEDAGOGY

When the *interact* component of interactive learning is taken to the next level, students are more involved in the educational process, and their participation is consequential.

With new media technologies, students become partners in shaping pedagogy. They influence how their own learning occurs by using technology more pervasively than in the process of interactive learning, described in the previous chapter of this book. With Web 2.0 tools and new media devices, students can collaborate with instructors to reinvent the fundamentals of classroom instruction: from establishing course flow, to organizing class modules, to determining how course materials are presented.

With this collaboration, there is a deeper involvement pedagogically than with conducting research and digesting course material during interactive learning, because students actually assist in developing methodologies for instruction, study, and research by using new media tools. This collaboration is made possible by the sophistication, intuitiveness, and ubiquity of new media. This partnership between professor and student results in the development of a new kind of pedagogy: *participatory pedagogy*. How does this happen? How does the collaboration materialize, and how do students not only engage with learning but also shape the educational process?

Let us begin with the assumption that *active* is more successful in education than *passive*. On a basic level, when students are actively involved with education, they learn more and better than students who do not participate as much. In addition, in our device-driven culture, students expect to be *part of* the educational process, more than they expect to just attend class and receive knowledge. After all, much of what undergraduate students do in their everyday lives with technology is participatory in nature. The undergraduate students involved in the case studies in this book are part of a young adult demographic that is so familiar and comfortable with the preponderance of technology—which connects them to information and each other at lightning fast speeds—that participating with technology is normal, commonplace. We might even describe it as second nature. Because students use technology so intuitively, *how* they interact with it while learning is important to the development of pedagogy in these technology-focused times. (The detailed demographic profile of this student population is presented later in this chapter.)

Tools and devices are no longer separate from learning. They are *part* of learning. New media as a discipline is no longer a means with which to educate students. New media is the foundation on which we educators build our pedagogy that will educate students. Students' interaction with technology in education, as well as their personal investment in technology—in terms of time, energy, and capital—significantly affect how students learn and, therefore, how we should develop our pedagogy. This is the underlying principle of participatory pedagogy.

Participatory pedagogy depends on an interactive learning environment in which students' hyper-involvement (a lot of interaction) results in alterations to conventional instructional methodologies and practices, as well as the creation of new ones. With Web 2.0 technologies, students and instructors form a unique partnership actively focused on Education 2.0—using technology to interact with information. Technology provides connectivity well beyond back-and-forth communication: professor to student to professor. It provides students the opportunity to co-create pedagogy with instructors. Participatory pedagogy begins with instructors requiring students to use Web 2.0 devices in class, and it concludes with observing how students receive the information with the devices and what they do with the information *in order to learn*.

The process of participatory pedagogy is broken down into five steps:

1) Instructors require students to use digital devices, such as smartphones, tablet computers, and laptops, in the classroom.
2) Instructors present course material and have students interact with the material on their devices.
3) Students interact with the material, exhibiting certain behaviors while using the technology.

4) Instructors observe and record these behaviors, which show how students address issues, solve problems, answer questions, contribute ideas, and ultimately learn.
5) Through these behaviors, instructors and students together create a new methodology for learning with devices.

It is important that in the classroom, these devices are used only for educational purposes—no logging on to social networks, surfing the web, answering e-mail, or unrelated texting. Instructors must strictly enforce this rule and establish other parameters that may be necessary to focus students' use of technology on education. By restricting the use of devices in the classroom to learning, not socializing or playing, instructors give new value to the devices, elevate the importance of technology in pedagogy, and more readily shape learning processes with them. By using technology in this specific way, students through their behavior make suggestions about the following:

1) How course material is delivered via devices.
2) The methods that instructors employ to teach and present educational content.
3) How students themselves learn using Web 2.0 tools.

My primary objective with participatory pedagogy is to streamline teaching processes and devise methodologies that create efficient and productive classroom experiences, for both my students and me. I implement participatory pedagogy when I create a new course that is focused on the use of technology or when I rework an existing course for a new grade level or a different student demographic—for example, reworking a new media fundamentals course originally taught to undergraduate students for graduate students who have years of professional experience. Getting feedback from the demographic taking the course is necessary to appropriately shape the course for that demographic.

In the first of six case studies presented in this book, I document the process and results of implementing participatory pedagogy.

Case Study No. 1: New Media Process and Product

The demographic of the student participants for this and all six case studies included in this book appears next, with one exception: the number of students in this first study was 12; in the other five case studies, it was 30.

Demographic Profile of Student Participants in Six Case Studies on the Integration of New Media into Classroom Instruction

Number of Students:	12 (Case Study No. 1); 30 (Case Studies Nos. 2–6)
Age Range:	18–22 years
Gender Ratio:	Male 65%; Female 35%
Occupation:	Student, Part-Time Employment
Home Location:	Midwest, United States 90%; Other 10%
Student Status:	Undergraduate

In this case study, "Case Study No. 1: New Media Process and Product," I worked with students in an advanced new media course of mine called Media Convergence Lab. In this course, students are required to collaborate with each other on the design and development of a media-rich site for a screen-based environment, such as a computer monitor, tablet computer, smartphone, or kiosk. Collaboration among students and the building of teams is important for two reasons:

1) There are many front- and back-end issues that need to be addressed for this complex project.
2) The class is comprised of students with disparate skill sets and design capabilities.

In order to ensure that students in the course can produce fully realized projects, I do the following:

1) I put students into teams, which collectively have the design skill and technological proficiency necessary to produce the required project(s).

2) I ask all students to bring to class a portable digital device—laptop, smartphone, or tablet computer—for use in class during the entire term of the course.
3) I begin instruction with the presentation of a design/production task that, on a small scale, represents the kind of project students will be required to create during the course. This is a kind of *mini project*, which is presented during one of the first class meetings. It provides a test case for the purpose of assessing students' capabilities with technology and evaluating students' process using technology.
4) I observe the process each student follows with the mini project, in terms of how individuals use their devices to meet the objectives of the exercise I have given them.
5) I note and record these processes.

I am not so much interested in what students produce as I am in *how* they produce it—how they use technology to implement process. I observe students using the devices as they complete the mini project. I observe how they individually create strategies for accomplishing tasks, such as designing the interface and writing the back-end code. And, I observe how they go about producing the project: integrating the design and the code, managing a list of deliverables, and bringing the mini project to completion.

For the purpose of this case study, I create two distinct categories: *process* and *product*. How students devise and execute their strategy for completing the mini project is *process*. What results from students' creation and integration of front-end design and back-end code is *product*. I rate all student teams in both categories. Everything that students do to engage in process and create product is done on their portable devices:

1) In the case of a student with a smartphone, the notepad feature on the phone provides a place to document ideas and take notes when discussing creative and technical options with teammates.
2) In the case of a student with a tablet computer, a software tool called Photoshop Touch, by Adobe—a streamlined version of the industry-standard graphics editor Photoshop, developed specifically for touch-screen tablets—provides a place to create quick and preliminary design concepts with the swipe of a finger.
3) In the case of a student with a laptop, a text editor or word processing program, such as Microsoft Word, provides a space to write

preliminary code for the back-end component and the ability to save it in a file format that can easily be opened by other teammates on various other devices.

4) In the case of another student on a tablet computer, who takes on the role of project manager, an online spreadsheet from Google Docs provides a place to organize all team members' tasks in terms of a) the project's objective, b) the intended result, c) the timeline, d) the distribution of information, and e) the various technological devices used to produce the mini project.

As students work, I observe that—no matter what the initial objectives or intended results of the projects—the devices students use are integral to how they engage in process and create product. The actual device determines *how* individual students brainstorm with one another and what they produce collectively in their teams. More importantly, the use of the devices determines and defines students' collaborative process. I observe *how* students create product—design, code, and so forth—in a process that is dependent on and necessitated by certain functional capabilities of their technologies, both hardware and software. And, I observe *how* the process of creating on a portable device determines the substance or non-substance of the resulting product, whether the product is a list of strategic goals, design composite, code template, team organization chart, or even an iteration of the complete project.

The upshot is that the tool becomes an extension of each student's thought process, or psyche. When new media tools are so integrated with learning, some psychologists would refer to technology as *cognitive prosthetics*: tools that make students better thinkers. The notion that tools become an extension of students' psyches and, therefore, are integral to their decision-making and learning processes, is key to participatory pedagogy. It is a notion put into context by Professor Andy Clark of Edinburgh University, who in a *New York Times* online commentary entitled "Out of Our Brains," wrote the following:

> As our information-processing technologies improve and become better and better adapted to fit the niche provided by the biological brain, they become more like

> cognitive prosthetics: non-biological circuits that come to function as parts of the material underpinnings of minds like ours.[8]

Because new media tools and devices are so integrated in our life experiences, they function as parts of our mind. We solve problems and make decisions with tools and devices by using them. Certainly there is the tactile re-reality of working hands-on with technology, to make the tools and devices do what we want them to do. But when our knowledge of the tools and devices is complete, our command of them full, our comfort level with them high, and our utilization of them frequent, they become "underpinnings of our minds." On some level, what we create with tools and devices is an extension of our will.

All that might sound highfalutin, but it makes sense in new media. And it can be applied to pedagogy, or in this instance participatory pedagogy. As I observed students working with devices to accomplish their goal of completing the mini project, I saw that they were *deciding by using*. Students' decisions about how they would go about completing the project were made once they chose the devices with which they would work. In this case study, I needed to make sure that each team of students had among them individual members with enough design skill, technological proficiency, motivation, and creativity to make the project happen. I was able to ascertain which students would work best together by which technologies they chose to work with—smartphone, tablet computer, laptop, and so forth. Students instinctually chose devices with which they had a level of proficiency. However, what really mattered in this case study was *how* they worked with their chosen devices to fulfill the objectives of the mini project, *how* their will was manifested when using the devices, and *how* they ultimately produced the project.

To recap, all students had a similar objective: make a mini project in the most efficient and creative way each team saw fit. This mini project would be a scaled-down version, or prototype, of a more complex convergent media project that would be created during the term of the course. Students approached the mini project in different ways, depending on their creative, technical, and managerial insight, their design and production capabilities,

---

[8] Andy Clark, "Out of Our Brains," published December 12, 2010, http://opinionator.blogs.nytimes.com/2010/12/12/out-of-our-brains/

their knowledge of new media concepts, and their experience working in a team environment.

Here are my findings at the conclusion of "Case Study No. 1: New Media Process and Product":

1) Design-focused students first worked on the front end, creating a user interface that had visual appeal (to them). Then the team took time to figure out how to support the interface with back-end technologies, or certain programming code that would make the interface function on students' individual devices.
2) Technology-focused students first worked on the back-end, creating a programming structure for a unique user experience in a screen-based environment. Users would access information by pressing buttons on the screen and swiping fingers across it. Then the team worked on designing an interface for the screen that would serve as a skin for the user experience the students coded.
3) Management-focused students first worked to organize a list of deliverables that documented specific tasks and deadlines. Second, the team distributed necessary information and digital files to team members via their devices. Then students oversaw the workflow of individual team projects. The team did not engage in the design or coding aspects of the mini project.

I then synthesized all that I observed during the development of this mini project (case study) and created a "Course Plan for Team Collaboration," which included a list of specific learning outcomes for the course—outcomes focused on the integration of new media technologies and students' behavior using devices. The course plan would work in tandem with the syllabus, already created, to guide students during the 10-week term of the course. (The university where I conducted this case study organized the academic year in quarters, as opposed to semesters, where a term would be closer to 15 weeks.) The syllabus for the course would provide students with standard course information: a list of topics that would be covered during the term, along with due dates and deadlines for assignments. The course plan would provide a pedagogic structure for the course: *how* devices, theories, and instruction would be integrated into the classroom experience, so that students, in teams, could follow process and create product.

The course plan for one of the teams appears next. This was a portion of the entire course plan, as information is presented here for one team involved in the mini project, not all teams. The complete course plan would include data for all teams, and it would document the design, programming, production, and management capabilities of all students in the course, organized in several teams.

Course Plan for Team Collaboration in Media Convergence Lab

Team No. 1

Role	*Device*	*Product*
Designer No. 1	Tablet computer	Design concept
Designer No. 2	Smartphone/Laptop	Final designs
Programmer	Tablet computer/Laptop	Code and widgets
Project manager	Smart phone/Tablet computer	Deliverables and deadlines

Team's Objective
The team's primary objective is to conceive, design, program, and produce an application for an international news magazine that will appear on a tablet computer with the Android operating system. The application's graphical interface will recall the design themes of the print version of the magazine. However, the new design will be streamlined in order to work within the limited real estate of the tablet's screen, versus the larger cover of the print magazine. The application's user experience will be intuitive but not predictable.

Team's Process
When creating the application, remember the concept of activity theory, which maintains that human beings interact with screen-based interfaces, or anything, for emotional or practical benefit. Make sure to focus your design and user experience so that it is primarily emotional *or* practical. This will influence your design choices—layout, color, visual balance, and so forth—as well as the complexity of the interaction design and coding. Use the devices required of team members, listed above, to do the following:

a. Communicate with the team.
b. Create the required project components: design, programming, and management.
c. Share project files with the team.
d. Track the progress of the project: producing deliverables and meeting deadlines.
e. Test the project's application thoroughly on a tablet computer that has the appropriate operating system. (In this case, Android.) During testing, make sure both aesthetic and technical issues are resolved.

Team's Product
A fully realized mobile application for a news magazine.

In summary, "Case Study No. 1: New Media Process and Product" provided a high level of student participation with Web 2.0 technologies that influenced how the instructor:

1) Formulated class modules.
2) Engaged students in process.
3) Presented course materials that guided students in the creation of product (a project).

A new methodology was born out of the case study. This was participatory pedagogy. This methodology would be implemented in the actual Media Convergence Lab course, which was separate from the case study and ran in a future term.

With participatory pedagogy, students took on an active role in shaping their own educational experiences. Students were key players in how they learned, and they were contributors to the actual methodologies that I, the instructor, would employ to teach them. Generally, students' participation, or actions, helped me to craft pedagogy: the principles and processes students need in order to learn. This information was valuable for the creation of new exercises and future courses. And it all began with cognitive prosthetics: tools that make students better thinkers. There is something both revolutionary and basic about this approach.

6

SOCIAL MEDIA AND COLLABORATIVE LEARNING

More and more, instructors are integrating social media into curriculum in order to engage students and prompt collaborative learning. The popularity of social media platforms such as Facebook, Twitter, and Google Plus in class sessions is gaining, as is the use of new technologies such as tablet computers and smartphones, on which social media can easily be accessed. Though more conventional technologies such as computers, laptops, and so forth, are still used in instruction, there is a shift away from these and toward handheld devices. Two reasons for this may be cost and ease of use. Because these technologies and platforms are relatively new in education, it is difficult to determine exactly why they are so popular, but research points to a definite trend in education: Social media use is on the rise.

In new media, we accept that *current* becomes *obsolete*. However, none of us can predict with accuracy how quickly this will happen. Advancements in new media will marginalize the use of current technologies. That seems certain. And there is a definite trend in technology turnover in education. The cycle of change in new media is a recurring issue in this book. The question educators must ask is this: Is the use of social media in education a trend, like some other new media, or are social media tools here to stay and, therefore, different?

According to a national survey, the number of faculty who use and support the use of social media in the classroom is substantial:

> Nearly two-thirds of all faculty have used social media during a class session, and 30% have posted content for students to view or read outside class. Over 40% of faculty have required students to read or view social media as part of a course assignment, and 20% have assigned students to comment on or post to social media sites. Online video is by far the most common type of social media used in class, posted outside class, or assigned to students to view, with 80% of faculty reporting some form of class use of online video.[9]

There are many examples of how social media tools and platforms are being used in the classroom. Here is a short list:

1) Instructors supplement lecture material with online readings, such as blog postings that are relevant to course instruction.
2) Instructors require students to respond to questions posed in class in online forums and to participate in discussion threads on social media websites that relate to topics covered in class.
3) Instructors ask students to contribute to class discussions via microblogs, such as Twitter, and then aggregate these comments for all students to view in a web browser, in and out of class.
4) Instructors require students to view videos on social media websites, such as YouTube or Vimeo, as an addendum to material presented in class.
5) Instructors post their own videos and audio podcasts on social media websites and have students watch and listen, in lieu of or in addition to in-class instruction.
6) Instructors have students post their work—writing, video, audio, and interactive—on social media websites.

Though there are statistics that show that social media have a substantial presence in the classroom, there are no data that prove unequivocally that social media are beneficial to learning. Similarly, there are no data that prove that social media are detrimental. We simply do not have enough years working with these specific tools, networks, and platforms to understand how they genuinely affect teaching and learning.

---

[9] Mike Moran, Jeff Seaman, and Hester Tinti-Kane, *Teaching, Learning, and Sharing: How Today's Higher Education Faculty Use Social Media* (Pearson Learning Solutions, New York, April 2011), 3.

We can surmise that teaching with social media is good, because, as mentioned earlier in this book, technology platforms, tools, and devices generally engage students, and students learn more easily when they are engaged. Following on the information presented previously in this book on interactive learning, we can also surmise that interactive learning is the ultimate way to engage students, because, after all, it is *interactive*. And *interactive* can evolve into a higher level of engagement and become *participatory*: Students become progressively more involved in the educational process, to the point of participating in the creation of pedagogy.

Therefore, it follows that when educating with new media, educators would focus on teaching with technologies that engage students at a high level. Social media platforms do this. Social media are all about connecting individuals to communities of people who have elected to become part of that network, because these individuals want to engage as much as possible and as often as possible with that network. The quality of the engagement does not seem to matter as much as the amount of engagement, which, for avid users of social media, is a lot. It is engagement at a high level.

For example, in the case of Facebook, a member of the social network becomes a *friend* with someone in the network, and this *friend* can then link the member to many other friends, or people, to engage with online. It is an ostensibly endless process with a potentially infinite number of connections, or friends, to be made. Becoming a friend means that you are networked to your friend's friends. The circle of friends grows. The amount of engagement increases. It goes on and on. (There is more about Facebook later in this chapter.)

In addition, social media, with their appealing graphical user interfaces, point-and-click media components, and intuitive user experiences, present an *ease* that can facilitate learning. There is a persuasive argument that facile learning makes education pleasurable: Students who learn easily will be contented. Contented students, motivated to learn and empowered by what they learn, are productive students. Productive students are successful students. In theory, social media can do all of this: make contented, productive, and successful students. But we do not know exactly how or why. We can only wonder at this point if the vast, expansive circle of friends has anything to do with it. There is, after all, power in numbers.

Still, we use social media in education a lot, despite the lack of formal research that shows whether or not these platforms work. Social media websites have become popular in the classroom for some educators, myself included, because of the collaborative nature and affordability of the platforms. Social media platforms and networks are easily available, and in most cases free, making social media in the classroom cost effective. Social media supports collaboration among students, as well as between students and teachers, in and out of the classroom. All that is required is a computing device, an Internet connection, and a browser, or a mobile application.

But as with any free service, there is a price to pay. Using social media effectively takes substantial effort. Incorporating social media into teaching means taking the time to examine technologies, use them, test them, and choose which ones support genuine collaboration. It is also important to filter out content presented on social media platforms—such as video, wiki postings, and blog commentary—that may be deemed inappropriate for instruction. For example, in my electronic media production courses that have a video component, I have students log on to YouTube and Vimeo and view videos from filmmakers, producers, and video artists from around the globe. However, before doing so, I pre-screen the videos and determine which ones support learning outcomes for the particular course module and are worthy of class viewing. Social video websites are a valuable resource, because in essence there is a library of clips from accomplished videographers, video artists, video educators, and video professionals accessible in a web browser for learning, ostensibly free of charge.

But there is also nonsense out there. Along with quality content on social media websites, there is a lot of unworthy content that is of little educational value. All this unworthy content—whether it amuses, shocks, stimulates, repulses, or just bores the viewing audience—is online all the time. This kind of content is referred to as *digital noise*. Unlike white noise, which exists as a constant background annoyance of little consequence, digital noise is in the foreground, in your face, and if it is not filtered it can have consequences in education. Students may believe erroneous information or choose to learn a skill by watching an unqualified person's online tutorial, which could teach them the wrong way to do something. Therefore, filtering social media content and other online content intended for use in instruction is essential.

After filtering, it is paramount to create synergy between the online content and the topic being taught in a course. Do not use online content arbitrarily. A mathematics instructor, for example, should not show videos satirizing politics. A digital media instructor should not send students to a blog about gardening. This seems obvious, but we all should avoid using social media just because they are there and jumping on the digital bandwagon just because we can. Like the requirement to use all tools with purpose, described in detail in the chapter, "Technology, Purpose, and Meaning," social media in the classroom are effective when used with real purpose. This takes effort. Instructors need to do the following:

1) Explore what platforms genuinely support a teaching agenda.
2) Determine when social media get integrated into teaching.
3) Decide how the platforms are used.

When defining the purpose for the use of social media in education, educators are redefining the role social media play in students' lives. We must clearly communicate to students where, when, why, and how we are using social media, so that the educational purpose is clear and use of the tools is effective. There is more on this later in "Case Study No. 2: Interacting with Literature on Facebook."

With social media we are reinventing how we teach. We are also reinventing how we communicate with students, for better or worse. Because social media are so prevalent in the classroom, electronic communication facilitated by social media platforms is usurping one-on-one, human communication, or at least seriously competing with it. This is exacerbated by the prevalence of devices students use to access the platforms. It is not uncommon to see students' heads buried in smartphones and tablet computers, with thumbs in a texting frenzy. All of this may become more of a concern, as more students rely on devices and social media platforms to communicate with instructors and with one another regarding a range of class matters: lectures, assignments, projects, examinations, timelines, collaborations, and so forth. With excessive use of social media platforms, we may be impeding open discourse in the classroom, marginalizing in-class discussions, and making verbal communication obsolete. Or, on the other hand, we may be opening up communication channels by providing a more comfortable, less intrusive, and streamlined way for students to answer questions, offer commentary, and contribute to class proceedings. Again, time and further use of these tools and technologies will tell.

Whether we are revolutionizing or hindering communication in the classroom, the issue of heightened student involvement because of social media remains real. This is good. This is helpful to students who are not naturally comfortable participating in class. In order to encourage more involvement among this type of student, some educators have integrated microblogs and social media websites into the classroom experience. These technologies give students who are shy, reticent, or just unwilling to comment or answer questions, a sense of security and an opportunity to add their voice in a non-threatening way.

Here is an example of how that might work. A student could contribute a thought to a classroom discussion via Twitter. The student would type the message into a device, hit send, and then the message would appear on a screen projected in the front of the class. The student's real identity may or may not be known, especially if the thought/contribution appears attached to a cryptic username. If the student's message is not the kind of contribution that the instructor would deem worthwhile, the student is saved from potential embarrassment, as long as the student remains anonymous and does not identify with the comment or username. If the contribution is meaningful, the instructor could make that known to the class, and the anonymous student could take credit.

Not only does this use of social media increase student involvement, it facilitates collaboration. One student's contribution could spur other students to contribute, and there potentially could be a snowball effect—an amassing of in-class contributions on a particular topic, which are posted on a big screen for all to see. This seems like a worthwhile use of social media, in particular Twitter. We may consider prompting contributions from students, aggregating the contributions in social media, and displaying them in class as a constructive use of technology in education. But it is not always.

There is a Faustian bargain with the use of certain technologies in education. Most educators want to push students beyond the boundaries that students have set for themselves, beyond their comfort zones. Educators want to get students to come out of their shells and contribute loud and clear with human voices. After all, students will need to communicate verbally in the workplace and communicate well. The classroom provides a training ground for such verbal communication, where students can acquire basic public speaking skills that are necessary for most professions. Social media may be compromising some students' development of these essential verbal communication skills.

The compromising effect that social media can have on interpersonal communication reaches beyond the student population. I know many people, professionals and scholars, who are fixated on the social media craze, who use Facebook and Twitter incessantly, and who are far less verbally adept than people who use social media in moderation. If we settle for a classroom environment that is comfortable and for communication that is streamlined and indirect, because of our use of social media, we may be doing students a disservice. Our job is to challenge students. Real challenges are uncomfortable and sometimes difficult. Social media provide easy communication, where quality and structure may be absent. Therefore, in terms of preparing students for their professions, communication with social media may be ineffective, incomplete, or both.

The bargain continues. We are mixing social media apples and oranges by having students use social media in a context that is not social. Traditionally, social media platforms have been used to make friends, spread news, garner support, upload photographs, view videos, post comments, and e-mail relatives. These platforms have been used for *social* purposes, in an environment that is fun, curious, and adventurous. We have taken the fun, curiosity, and adventure out of students' social media experiences by bringing the platforms into the classroom. There certainly is good intention here. One of our motives, as cited earlier, is to involve students who might otherwise be on the periphery of class participation—the shy ones. However, this is not fun. This is sensible. Have we educators really thought about this mixing of social media apples and oranges, this mixing of play and practicality? When we use social media in education, we use social media clinically, with the objectives of making classroom instruction more efficient, communication more effective, and the learning experience more productive. However, we do this with little or no formal redefining of the platforms for students. We need to redefine the use of social media for students first.

This redefining, or repurposing, of social media platforms and technologies—which outside of education are integrated seamlessly in everyday social interactions—requires thought, planning, and deliberation. We must be diligent and deliberate in the pedagogical application of these platforms, because any technologies meant for self-expression or casual communication, such as social networks, will be used for effective teaching differently than for effective living.

Our task is to make education complete with social media. Imposing social media on teaching and learning can be awkward and unproductive. Therefore, a set of rules for the use of these tools specifically for education is required. These rules, which determine how and when students will use the technologies and platforms in their learning, must be established at the beginning of a course, with the creation of the course syllabus. These rules should remain part of the pedagogy that an instructor develops for a course, or, in the case of participatory pedagogy, that the instructor and students create for the course. These rules will also need to be tweaked or rewritten when a course is reworked for the future, when more technologies and platforms get integrated into the course, or to accommodate changes in technology. However, the core reasons—the pedagogical objectives for using social media tools, technologies, and platforms—will remain the same.

In each of the case studies presented in this book, I created a set of rules that students needed to follow, in order to effectively use certain technologies and platforms required during instruction. The use of new media was integral to realizing the learning outcomes in each case study. However, the rules varied from case study to case study, as did the technologies and platforms and the objectives for using them. My core belief, that the tools should always serve a pedagogical purpose, remained in the forefront of my instruction, including the case study when I integrated the social network Facebook in my teaching.

Facebook is a social media platform that in some way defines what social media are and what processes are commonly accepted for online social interaction. The ubiquity of Facebook has turned global citizens into a community of digitally connected *friends*. With Facebook, we build community and create *friends* merely by joining the social network and participating in a series of clicks, postings, comments, and *likes*. The concept that we can click a button to request friendship with *people you may know* and that those *people you may know* can reciprocate by clicking a button to solidify the relationship is strange. It *is* strange. That said, clicking to create relationships does not seem to be going away, and neither does Facebook. So, let us seriously consider it for education. Beyond the bizarre protocol and instantaneous friendships that may develop, the social media platform has merit as an educational tool.

In a discipline like electronic media, students are eager to hone production skills. And while I view proficiency with technology and tools as sensible and necessary—proficiency will help students secure an entry-level position upon graduation—a preoccupation with the tools alone will not create value in their education. We want them to push buttons with purpose, in order to produce meaning, as discussed earlier in this book in the chapter "Technology, Purpose, and Meaning." And what better way to give button-pushing purpose than by beginning an assignment with the reading of a well-written text, whose subject matter resonates with students' intellectual faculties and has relevance to the course, and which has a reasonable length, so that the book can be read and pondered with enough time left in the term for students to learn the necessary course topics and production techniques required by the syllabus.

In the next case study, I chose to combine both literary and social media experiences by having students read a thought-provoking book and then take an examination on the book on Facebook—in class and in front of their *friends*.

Case Study No. 2: Interacting with Literature on Facebook

You must read a book with contemporary philosophical themes and take an examination on this book in class on Facebook. That was the directive given to a class of my new media students for this case study.

I required my students to read *Blink: The Power of Thinking Without Thinking* by Malcolm Gladwell, the *New Yorker* writer, who pens books about the kind of contemporary socio-cultural issues that I felt my students could relate to: understanding first impressions, the growth of social trends, and what causes success. You may recall that I used Gladwell's book in a research exercise documented earlier in this book, in the chapter "Interactive Learning." *Blink* takes on more significance in this case study. The book is about first impressions and gut instincts, and the value both can bring us when we act on them. The subtitle of his book, *The Power of Thinking Without Thinking*, indicates this. Gladwell's insight into contemporary issues has relevance to an undergraduate student's educational journey.

> We believe that we are always better off gathering as much information as possible and spending as much time as possible in deliberation .... But there are moments ... when our snap judgments and first impressions can offer a much better means of making sense of the world.[10]

Trusting instincts can prompt students to make the right decisions in pursuing careers upon graduation. And in a discipline like electronic media, go-

[10] Gladwell, *Blink: The Power of Thinking Without Thinking*, 13–14.

ing with gut feelings will empower students to try new things, including learning new tools and technologies, without the fear of failure.

From the beginning of this case study, there was more at stake for my students than just reading a book. What was really at stake was using technology in education meaningfully and "determining the value-creating capabilities of the technology."[11] The technology was a social network, Facebook. Facebook's value in this exercise was to provide a platform for an interactive, online examination that students would be given on what they read in *Blink*. Here are the instructions students were given, and they followed:

1) Read the book.
2) Distill the book's message.
3) Understand the book's theme.
4) Reflect on the book's subject matter.
5) Set aside the book.
6) Take an examination on the book on a social network.

It was conventional practice to have students read a book, ponder it, and then take an examination on what they read. However, it was unconventional practice to assess how well students knew the material and how much information they retained from the book with a social network. Assessment with Facebook was the focus of this case study.

Having read the book, students were required to follow the steps listed next before attending class to take the examination:

1) *Create a Facebook account specifically for class use.* Creating a Facebook account is a streamlined process, as easy as creating a membership account for any online platform or community. It was imperative that students did not use their personal Facebook accounts, as I did not want students to expose their classmates or me to information or media posted previously on these accounts, which might be of a sensitive nature.
2) *Post a unique photograph.* Facebook members have the option of posting a photograph (in the form of a thumbnail image) on their accounts to visually identify their profiles. I required students to post a

[11] Rushkoff, *Program or Be Programmed*, 13.

thumbnail image, but one that was not of a true likeness. Students would create a representative image, an icon, which I, and their class peers who would view all examinations online, could associate with individual student answers. This icon became a unique identifier. Students were asked to not use their real likeness, because of privacy matters.

3) *Friend the professor.* Like my students, I created a special Facebook account for this exercise. All students were required to *friend* me before the start of the examination.

In return, I, the instructor, confirmed my students as *friends*. This step was essential, because, once I was a *friend* with every student in this particular class, I would have an online connection to all of them for the posting of examination questions en masse. With all the necessary Facebook accounts created and Facebook friendships formed, students were ready to take the examination, online and in-class. On the day of the examination, students were instructed to:

1) Take seats in the classroom, which was equipped with a wireless Internet connection.
2) Open and turn on their laptops.
3) Log in to their Facebook accounts.

I did the same and projected my Facebook account on a large screen in the front of the class. I accessed my Facebook wall, a section of my Facebook profile where students' answers to my questions would be posted, and the examination began.

The process would unfold over three phases: I, II, and III.

Examination Process: Phase I

In the first phase of the examination, conducted with 15 of the 30 total students, individual students would answer questions by posting to my Facebook wall. Via Facebook, other students would choose, or *like*, the best and most complete answer and explain why they chose that answer. The process occurred as follows:

1) I selected three students to respond to the same question, which I read aloud.

   Students' names came from the official class sign-in sheet, which documented the students who were present. This provided an arbitrary order in which to call on students—the order in which they signed in, as opposed to alphabetical order or reverse alphabetical order. This arbitrariness added fairness to the examination, in that specific students would not be "picked on" to answer particularly difficult or easy questions.

2) Each of the three students had 1.5 minutes to think about the answer and post it on my Facebook wall.

   It was important to time students, because all students would need to participate in the examination before the end of class, which lasted 1 hour and 20 minutes. This time constraint added to fairness. As answers were posted to Facebook, I intermittently refreshed my Internet browser, so that the answers would appear in a timely manner.

3) I selected three additional students to choose the best, most complete answer of the three answers provided.

   Students did this by choosing to *like* a particular answer. This happens on Facebook by clicking on the *like* hyperlink that appears under each answer (entry). *Likes* are totaled (on the back-end) and the total number is posted online aside a thumbs-up icon. The total number of *likes* appeared under respective answers (entries), along with the student's Facebook account name, as well as any comment that student might post. Using the *like* feature, and being able to associate a *like* with an individual student, would provide me with necessary information when grading this examination. Students were being graded on both the questions they answered directly, as well as the correct answers they supported, or *liked*. Similar to the process in step number 1, these additional three students were selected from the official sign-in sheet. On the sheet, they were the next three listed.

4) The process was repeated until all 15 students (of the 30) had answered a question directly and also chosen to *like* (support) or *not like* (not support) the answers provided by three other students.

By providing all students in Phase I the opportunity to answer directly and also to *like* (support) answers provided by other students in the class, I created a framework for assessment. I would revisit this Facebook page after the examination to reread the answers and comments and to tally the *likes*, all for the purpose of grading students.

Ancillary to the framework created for assessment, I observed that an interesting kind of community, a community of peer support, was created around answers perceived to be correct: the more students who thought an answer was correct, the larger the number of *likes*, the larger the community.

Examples of Questions and Answers from Phase I

Before describing Phase II of the examination process, I would like to present from Phase I an example of the question-and-answer process. Following is 1 examination question, the answers provided by three students, and the three comments posted about those answers by students who *liked* the answer.

Question No. 1
In the chapter "The Theory of Thin Slices: How a Little Bit of Knowledge Goes a Long Way," a SPAFF code is employed in an experiment. In this context, what does SPAFF stand for and what is predicted by analyzing the results of this experiment?

Student Answer No. 1
SPAFF refures [*sic*] to the experiment which was conducted by John Gottman. This experiment involved a numerical system which quantified the emotional reactions of humans. Gottman specifically focused on the emotional exchange between married couples.

Student Comments on Answer No. 1:
None

Student Answer No. 2
SPAFF is the process of looking at someone's facial expressions and assigning numbers to those emotions. With these numbers, heart rate increases and other things can be determined. The example in the book had Gottman choose whether or not the couple he was looking at would remain married or not.

Student Comments on Answer No. 2
I think this process is very cool, but I think a lot more than just facial expressions goes into whether or not a marriage is going to last.

It was pretty interesting that his experiment ended up having such a high success rate, but I suppose those hidden emotions he's trying to pull out are very important to the length of the relationships.

Student Answer No. 3
SPAFF is a coding system that has twenty seperate [*sic*] categories corresponding to every emotion that a married couple express during a conversation. Disgust = 1 contempt = 2 anger = 7 defensiveness = 10 whining = 11 sadness = 12 stonewalling = 13 neutral = 14. From the SPAFF coding system results Gottman has proven whether the couple will still be remained married 15 years later.

Student Comment on Answer No. 3
The SPAFF system shows how there is an enormous amount of data that is applied to a quick judgment such as thin-slicing. It is another example that Gladwell uses to show how a system such as SPAFF can contain so much information but really generates a quick simple action, emotion, predicton [*sic*], or mind-reading task.

Correct Answer to Question No. 1
Upon completion of the examination, students were provided with the answers. Here is the correct answer to question No. 1: SPAFF

stands for Special Affect. It is a system that determines if a couple will be together in 15 years.

It is interesting to note how students entered their answers: with partial sentences, inconsistent punctuation, and cryptic spelling. It seems that the preponderance of social media in our everyday lives, and the way we communicate with them, has made an impact, perhaps a lasting one, on the way students spell.

Examination Process: Phase II

In the second phase of the examination, conducted with the remaining 15 of the 30 total students, 1 student would answer a question by posting to my Facebook wall. All other students (14) would chose to *like* the answer or not, depending on if they thought it was correct. Phase II differed from Phase I in that 1 student, not three, was required to answer each question, and all remaining students, not just three students, would choose to *like* the answer to the question, in essence voting for the correct answer. This was repeated until all 15 students had the opportunity to answer a question.

In this phase my objective was to provide students more choice in terms of right and wrong—15 total questions as opposed to 5 in Phase I. Having more questions and answers provided students with a deeper analytical experience and more opportunity for critical thinking. There was more information for students to ponder and assess. This phase was more challenging for students. Also, the process of *liking* was altered during this phase, in that students would not add commentary.

This phase provided a different dynamic for community building among students. The formation of community was better defined in Phase II than in Phase I. In both phases the size of community was measured in *likes*—the more *likes*, the bigger the community. However, in Phase II it was easier to aggregate a person's community, and community, in terms of the number of students who supported another student, was more apparent. Considering all of this, the community building aspect was more consequential in Phase II than in Phase I. It was more indicative of the process of expanding a social network.

Examination Process: Phase III

The third phase happened after students had answered all questions, the examination was over, and the class had ended. This is when I calculated grades. It was an interesting exercise for me. For a large part of my career as a professional educator, I had never had the opportunity to log on to a social network, access an account created specifically for a course examination, and review information posted on the website for the purpose of grading students. While at first it seemed odd—again a kind of mixing social media apples and oranges—I soon found the information easy to access, interpret, and grade. Facebook provided an online archive of the questions, answers, and comments from the examination that was complete.

It should be noted, however, that if I had not required students to create unique Facebook accounts specifically for the examination, assessment would have been challenging. The examination data I was revisiting would have been, over time, interspersed with other streams of data that were fed into the network by students' friends outside of class. This would have made the grading process difficult.

Assessment

I reflected on the experience of conducting an examination with a social media platform. Certainly, it was different. I wondered if it were right for education. Did Facebook serve my pedagogical objectives of involving students with content first and technology second? It did.

7

BACKCHANNELS AND MULTITASKING

Educators ask and sometimes require students to go online for educational purposes: as a component of a learning module, the focus of a class exercise, a method of research, and/or a means to engage with other students. Human interaction with online platforms and digital devices is so prevalent in education today that it is common for students to assume that technology, not just devices but entire systems, will be a part of their studies. Moreover, educators and the institutions where they teach are building technology-based systems specifically to facilitate student interaction in the classroom. It seems that these systems can do positive things for education. These systems can engage students in virtual learning environments, enhance communication between students and teachers, and support peer-to-peer collaboration.

In the classroom, these systems can be in the foreground, the background, or both. When in the foreground, these systems constitute a main communication stream and support primary communication activities, such as surfing the web, answering e-mail, reading a magazine on a tablet computer, and engaging with a mobile application. A main communication stream involves human-to-human exchanges—people talking to and with one another. A main communication stream in the classroom is typically one of the following:

1) A teacher instructing a class of students.
2) A teacher posing questions and students answering them.
3) A debate among students on a particular topic.
4) Students offering commentary on material presented in class.

When in the background, these systems are called *backchannels*, and, though they are subordinate to any main communication stream, they can be just as important in teaching. A backchannel creates a secondary way of communicating. Usually, a backchannel is technology-based, but not always. The backchannel could be as simple as the intermittent use of a microblog like Twitter integrated into course activities. Or, the backchannel could be the more widespread use of devices and equipment, such as mobile devices, tablet computers, and the applications that can be stored on them, in everyday teaching. Or, the backchannel could be more visible, such as the projection of lecture information and/or learning materials on a large screen for all people in the class to read. These are all examples of simple backchannels.

Backchannels can also be complex. The backchannel could be a proprietary server-based system that a university develops on its own that allows students to participate in class virtually, by posting comments to a web platform, which then could be projected on a screen or viewed on myriad devices at any time. Or, the backchannel could be a sophisticated 3-D online community, which requires members to represent themselves as avatars and inhabit a strange world that has its own currency, code of ethics, and set of laws.

The basic point with backchannels is that there is more than one channel of communication open at any one time, whether simple or complex. There is the main channel and one or more backchannels used either simultaneously or in tandem with the main channel. In either case, instructors and students have options, in terms of the means they use to communicate with each other. It would seem that with the many systems available to communicate, technologically and otherwise, instructors and students would be able to deliver messages, as well as share thoughts, comments, and suggestions in the classroom with ease and frequency. However, technology systems can also get in the way of education by muddling pedagogic processes. A preponderance of channel options could create a situation that is as detrimental as it is beneficial to the teaching/learning process. More channels may not necessarily make for better communication of ideas in the classroom. Suffice it to say, there are both pros and cons that educators need to consider before integrating backchannels into classroom instruction.

There are progressive educators, who applaud the use of backchannels and the concept of dual or multiple communication streams. These progressives would argue that with backchannels, classroom conversations, debates, and question-and-answer sessions are richer—if not for the content, then for the fact that students, who once may have been inhibited to participate, now have a voice. Albeit, their voice would be a digital one, likely represented as computer-generated text on a screen or as an avatar in a 3-D modeled environment. On the other hand, there are skeptics who denounce backchannels as tools that hinder the educational process by creating a multitasking environment in the classroom that undermines student productivity. These skeptics view backchannels as distractions.

Whether backchannels enhance or diminish the learning experience is certainly open to debate, and I encourage professional educators to be active in the debate. Should we have a single, focused communication stream in the classroom, where students are required to tune into a single main message being communicated by the professor—thus making for deliberate and focused communication of the message? Or, should we introduce technologies that splinter communication, where varied and different ways of connecting with students are used—thus making no single stream of communication more important than another and perhaps making the educational environment more interesting and exciting for students and instructors?

Before setting out to determine which one was the better option for my students, and thereby defining my use of backchannels in the classroom, I wanted to examine something more basic, which, I believe, is at the core of the debate over backchannels—*multitasking*. Generally, multitasking involves engaging in one or more activities or undertaking one or more tasks at the same time. In the classroom, multitasking can cover many tasks and take many forms, but, for the purpose of this discussion about backchannels, I focus on multitasking with technology.

It is not uncommon for students to answer a cell phone or text on a smartphone in class, even if the use of the device is not connected to interactive learning or the creation of participatory pedagogy. Students, and some professors for that matter, just want to be connected as much and as often as possible. Why not allow technologies, tools, and devices in the classroom and create a situation where students can multitask? Texting on a smartphone while listening to a lecture would be an example of in-class multitasking, and if both tasks could be completed with positive results—for example, the text message was composed and sent, *and* the lecture was understood and absorbed—there would be a lot of productive students in classrooms around the globe. Students would be engaged communicators with high grades. Multitasking in the classroom would be good—in theory. In practice, multitasking is infeasible.

Research has shown that multitasking, which we require students to do when we integrate backchannels into their classroom experience, is not favorable to productive learning. Specific studies of multitasking with digital devices, which is the crux of the debate over backchannels in the classroom, have been conducted and the research published. The research concludes that, generally, multitasking has serious shortcomings. In an April 2011 *New York Times* article on multitasking in our technology-driven lives, the opening sentence read as follows:

> A growing body of research shows that juggling many tasks, as so many people do in this technological era, can divide attention and hurt learning and performance.[12]

Dr. Adam Gazzaley, a neurologist at the University of California at San Francisco, was quoted in the same article, and he commented as follows:

> This issue is growing in scope and societal relevance as multitasking is being fed by a dramatic increase in the accessibility and variety of electronic media.[13]

In another *New York Times* article about students and multitasking, we learned that:

> Students have always faced distractions and time-wasters. But computers and cell phones, and the constant stream of stimuli they offer, pose a profound new challenge to focusing and learning.[14]

And, in yet another New *York Times* article, documenting collaborative research conducted by *New York Times'* writers and brain scientists, there was the following information:

> While many people say multitasking makes them more productive, research shows otherwise. Heavy multitaskers actually have more trouble focusing and shutting out

---

[12] Matt Richtel, "Multitasking Takes Toll on Memory, Study Finds," published April 11, 2011, http://bits.blogs.nytimes.com/2011/04/11/multitasking-takes-toll-on-memory-study-finds/?scp=1&sq=multitasking&st=cse

[13] Richtel, "Multitasking Takes Toll on Memory, Study Finds."

[14] Matt Richtel, "Growing Up Digital, Wired for Distraction," published November 21, 2010, http://www.nytimes.com/2010/11/21/technology/21brain.html?ref=yourbrainoncomputers

irrelevant information, scientists say, and they experience more stress. And scientists are discovering that even after the multitasking ends, fractured thinking and lack of focus persist.[15]

The research cited previously is just some of the trove of information available to us that makes the case against multitasking in the classroom, and in daily life for that matter. However, I venture to say that proponents of multitasking could find information and research that supports the use of backchannels in the classroom. For example, there are platforms such as Today's Meet that are available for backchannels, where teachers can set up a virtual room in which students can communicate, and Second Life, a 3-D environment in which students interact as avatars. Both of these platforms have been used extensively as backchannels by educators, who believed that the platforms added a meaningful way for students to communicate and interact in the classroom. However, technology comes and goes, and Second Life, once touted as a force in the reinvention of education with technology, is hardly as popular in instruction as it was years ago. When asked about Second Life recently, most of my students had never heard of it.

Still, technology in education is au courant, and in our new media-centric culture, multitasking with technology in education is sometimes depicted as chic and hip. Photographs, video clips, and visuals all around us—including on the World Wide Web, on television, and in film—show students in classrooms creating content on tablet computers, *while* texting with mobile devices, *while* looking at projections of course material on large screens. These visuals, which document a productive, cheerful environment of backchannels, are persuasive, perhaps even propagandist. They promote a kind of education that is contemporary, forward thinking, and tech-savvy. It is what trend-focused academics and career-driven administrators at institutions of higher learning want education to look like. This kind of education sells multitasking as a way to achieve academic success.

So, why are the creation of backchannels and the introduction of tools, devices, and virtual environments in the classroom popular in some educational circles and not others? I wonder if they *are* popular or if they just get an inordinate amount of attention and press. But the facts are the facts. We know that multitasking compromises focus, attention, and productivity in the

[15] Matt Richtel, "Attached to Technology and Paying a Price," published June 6, 2010, http://www.nytimes.com/2010/06/07/technology/07brain.html?ref=yourbrainoncomputers&pagewanted=2

classroom. The spin culture of Madison Avenue and Silicon Valley that sells multitasking as absolutely right for education cannot be the guiding force for educators, without further investigation. The facts do not support a revolution in education led by multitasking—*yet*.

Technology, when used properly and purposefully, is powerful in education. But let us introduce technology into the classroom as needed, as it supports pedagogic endeavors and helps meet learning outcomes, not because it is trendy. Hip and current, though appealing, are not necessarily the qualities of solid education. Let us use technology to connect students to the material we teach, in a visceral and sustaining way. Let us avoid using technology simply because it is there. Let us use technology because it brings meaning to teaching.

Bringing meaning to all aspects of teaching is my primary objective as a professor. Near the beginning of this book, I discussed the evolution of purpose into meaning when using new media tools in education. Now, I add multitasking to the discussion, and I pose the following question: Can multitasking, though shown in prior research to compromise students' focus in the classroom, be used in a new way—a yet untried way—that will genuinely connect students to course material, support pedagogic endeavors, and promote meaning in learning? After reading the *New York Times* research, cited earlier, two phrases resonated with me:

1) "Constant stream of stimuli."
2) "Fractured thinking and lack of focus persist."

I wanted to see for myself if, in a classroom where a "constant stream of stimuli" persisted, students would be left in a state of "fractured thinking and lack of focus"—after the class had ended and the technology was put away. So, I created a unique educational situation for the use of multitasking. I set out to conduct an informal experiment, not a formal case study, in which I would observe students during and after a class of multitasking, and assess what effect, if any, multitasking had on student performance. I chose one particular course, Advanced Interactive Media. Students enter this course at the beginning of the term with established skills and knowledge of certain tools and technologies, as mandated by a list of prerequisites, but no student knows *all* the tools and technologies taught in the semester-long course. So, there is a learning curve on some level for all students.

For this experiment, I created a single three-hour class with as much stimuli—aural, visual, and cerebral—as I could provide for my students. I crafted an assignment that would require students to work with a number of media types and new media technologies. The assignment would be both creatively and technically challenging, and it would have to be produced in a short period of time. The project, which students would need to complete by the end of a single class session, was a web-based media player that would need to play both audio and video files. The player would be built using a scripting language of the students' choice. The two most likely choices were jQuery and ActionScript, as these were the languages with which students had the most experience.

The exercise began with the accumulation of assets—audio and video files—from an online archive of files available to students. It then continued with the creation of the media player itself, both front-end interface and back-end functionality. To provide added visual stimuli as students began to work, I turned on the projector and began projecting, on a large screen in front of the lab, images from multimedia projects that students had created in a previous course of mine. These images documented projects unrelated to the media player students were currently building. The size of the projected images on the screen was so large that all students in the moderately sized lab space would have the images in view, at least peripherally. For added aural stimuli, I played music of various genres at a loud volume over a set of sophisticated speakers in the classroom. (The sound on these speakers was rich and robust.) The projected images and the music added stimuli that students had not experienced before in this course and were not used to having around them while they worked. Few students asked why I was supplying the aural and visual stimuli, which was interesting to me. For those who did seek an answer, I responded by saying that the audio and video playing around them were there to create a media-rich working environment but that students were not required to focus on either.

I observed students working to see if the "stream of stimuli" would create "fractured thinking." My criteria for determining "fractured thinking" was 1) whether or not students completed the project in the designated time, less than three hours; and 2) the quality of the work students produced. For the most part, students worked slower than usual. Initially, I attributed this lack of productivity to students' lack of focus, due to the increase in stimuli around them. It seemed that the newly introduced sounds and sights in the room *were* a distraction, even though I instructed students not to focus on them.

But this was not true for all students. There were two students who wore headphones and listened to personal audio files they brought to class for their projects. These two students were working on their projects more industriously and faster than the other students. They did not seem distracted like the other students. These two seemed focused on the work. Other students, who did not have headsets, had to listen to the audio files they chose for the assignment, from the available archive, on their computer station's built-in speakers. The sound from these speakers was not sophisticated, quite tinny actually, and it played in the open classroom. The audio coming from built-in computer speakers competed, in terms of volume and clarity, with the music coming from the classroom speakers, as well as the music coming from all the other students' computers' built-in speakers.

In terms of potential distraction from the large visuals being projected, certainly the students wearing headsets could see the images on the large screen in their peripheral view, like the other students could. But these images did not appear to distract many students at all, with or without headsets. I surmised that there is so much visual stimuli in the daily lives of undergraduate students in electronic media that students are either immune to the affect of this kind of visual annoyance or they have successfully learned to filter it out of their consciousness and, therefore, ignore it. Perhaps to some students, the visuals were not distracting or annoying at all; they were pleasant. It was the conflicting audio that created distraction. And the students wearing headsets and listening to their own audio seemed free of this distraction.

At the end of the three-hour class, the two students wearing headsets completed their projects, whereas a majority of the students exposed to the barrage of external sounds did not. The work from the two students wearing headsets was not necessarily more polished or better than other students' work, but they finished the projects. Again, students in this advanced class possessed varying skill sets and levels of expertise, so there was equity among students in terms of the quality of work each of them was capable of producing.

Based on my observations, I could draw the conclusion that the lower the level distraction, with audio in particular, the higher the amount of productivity with the project. But I would do so in the context of this being an informal experiment, not a formal case study. One bit of information I will add is that *choice* may have played a role in students' ability to focus and multitask with successful results. In addition to filtering out the extraneous audio that played on the classroom speakers as a constant stream of stimuli, the two students wearing headsets *chose* the audio they were listening to. They listened to *their* music, not that available from the archive. Therefore, I would conclude that *choosing* a stimulus minimizes its level of distraction.

I next focused on the other students, who did not choose their own music to listen to and did not finish their projects. To determine if a state of "fractured thinking and lack of focus" persisted in these students—who ostensibly had been adversely affected by the in-class stimuli during this experiment—I asked them to take the first hour of the next class to complete their assignments. I observed that most of these students did not pick up where they left off. A majority of these students rethought a number of the assets—audio and video files—they had chosen for their media players, and they selected new files. By reconsidering their original choices, they took a step or two back in the process of creating the project. These students spent time in this second class doing work that should have been done in the first class. This could be construed as a lack of focus, overall.

The idea of choice in multitasking is interesting to me, as I am a proponent of empowering students with choice in various aspects of their education, all within the framework of a solid pedagogy. I am reminded of what the Greek philosopher Aristotle wrote: Every choice is thought to aim at some good. With further observation and research on multitasking in the classroom, specifically in a new media environment, which I plan to conduct in the future, there may be two categories that will evolve and allow us to continue the discussion of multitasking at the next level: 1) multitasking with choice and 2) multitasking without choice. Comparing research done in both of these areas will provide additional insight into multitasking and backchannels.

One of the themes running throughout this book is this: If there is a solid pedagogic reason for using a tool or technology, use it. Some new media technologies that are used as backchannels—that may also be used to create an environment of multitasking—can be effective when they support a main communication stream. What is most important in education is that backchannels, if used, create a meaningful connection between students and course material, and the environment we teach in is unfractured and focused.

8

MICROBLOGGING IN THE CLASSROOM

With the social media platform and microblog Twitter, users have 140 characters to communicate individual messages, called tweets. Conventional media observer and author Marshall McLuhan coined the phrase "the medium is the message"[16] in his book of a similar but not identical title, *The Medium Is the Massage*. (The typo in the book's title is the result, ostensibly, of a printing error.) It follows then, using the wisdom of McLuhan, that the medium Twitter is a message: 140 characters of message. Because this message is communicated in the classroom, it is an instructional medium, and the medium/message should be one of substance, one that is relevant. But it seems that any medium, or tool, that allows for only 140 characters of expression at a given time would not provide for a worthwhile exchange of ideas in the classroom. If relevant and meaningful discourse is measured by numbers of characters, words, or sentences, then microblogs could not support relevant and meaningful discourse. Microblogs limit discourse. They are, after all, *micro*. So let us ask this: Can Twitter and other microblogs be relevant in classroom instruction, despite their inherent and obvious limitation on discourse?

A *serious* educator would put forward that the message brevity inherent in Twitter is indeed a limitation that undermines the communication of message. Serious educators want students to express themselves in full sentences, with proper grammar and adherence to the rules of punctuation. With 140

[16] Marshall McLuhan, *The Medium Is the Massage: An Inventory of Effects* (Bantam Books, New York, 1967), 10.

characters, punctuation *has* to be comprised. On Twitter characters are sacred. When tweeting—the verb that describes the process of drafting and posting a Twitter message, a tweet—you do not want to waste a character on a period or a comma. You need to express as much as you can, as creatively and directly as you can, with the 140 characters. The people who are avid about posting tweets—called either tweeters or Twitterers—treat the selection and use of characters to communicate message as an art form almost. In visual design, the choices you make about color and layout are paramount to what is communicated. On Twitter, your selection of characters and *what you do* with the sacred 140 characters are paramount and make your tweet impressive, or not.

A large part of why people tweet is to impress a general audience, professional clientele, and other tweeters. If you impress people, they will elect to *follow* you, meaning they will receive your tweets automatically. Having a following is essential to building community in the blogosphere, micro or macro. Your following is a collective of like-minded people who want to receive your messaging and perhaps other agenda you may be inclined to distribute via the Internet. Once you have created a community that thrives on your messaging, you can become a blogosphere leader, a digital celebrity, and a cultural guru. The guru status you achieve is a phenomenon that, I believe, can only happen in the anonymous world of Internet blogging. In the microblogging world of Twitter, you rise to the top because of what you do with 140 characters—how uniquely and succinctly you express yourself. Guru for 140 characters, *that* is impressive. And, you don't have to worry about punctuation.

But the blogosphere is not the classroom. The rules in the classroom, in terms of how a student should properly communicate, are different. Historically, these rules have supported a structured approach to language, syntax, grammar, and punctuation. If a student's grammar or punctuation were a mess, the *serious* educator would consider the student's communication inferior. Following this thinking, Twitter would be an inferior communication medium. But Twitter has revolutionized how people communicate to the world. It has millions of users, and it creates Internet superstars—in academia, business, and other areas. Can a technology platform with so much global impact be inferior, or conversely, can a tool that promotes bad spelling be serious?

Perhaps we need to redefine *inferior*. With new media an integral part of education, everything old is not necessarily new again. There are rules for in-class communication, ways of doing things, that are tried and true, and processes that have been standard for decades. It may very well be time to rethink some of these. We have a new medium, the Internet, which is driving how we communicate in unique ways. We need to shape communication methodologies, in and out of education, that are equally unique. And, we must continue to do so with the invention of new communication technologies. This is part of the process of reinvention that has always persisted in the world of media, new and old.

In the middle part of the 20th century, television was a force that put the elaborate prose of literary masters like Melville, Poe, and Dickens in a new context. Television gave society a more casual approach to language, grammar, and punctuation. Television was a medium (and a technology) widely used in education. Because of this, the less-than-formal language television provided eventually found its way into the vernacular and into the classroom. And, it was accepted by educators, not at the expense of the written language of Melville, Poe, and Dickens, but in addition to their language. That vernacular, that kind of new communication, was not inferior, like tweeting and texting are not inferior today. That new communication was just different, like microblogging is different.

Perhaps we need to redefine *serious*, at least in terms of communication in the classroom. Let us reinvent conventional paradigms, so that tweeting, texting, and future means of communication constitute new media that serve pedagogy, as radio and television were new media that served pedagogy. The medium that gives you 140 characters to express an idea *can* be a serious medium and a substantial message. Let us redefine what communication in the classroom with new media technologies means. Let us begin by answering the following question: Can the medium/message called microblogging, with its lack of proper punctuation and grammar, be appropriate for the classroom instruction?

The microblog can be a useful way to communicate ideas in the classroom, if the educator works *with* its limitations and puts the platform into context for students. We cannot pretend that tweeting is bad English. It is not. It is *tweeting*. As illustrated in the following case study, "Case Study No. 3: Engaging Students with Twitter," when the 140 characters are reused and repurposed in an educational process that is structured and deliberate, they add value to communication in the classroom. When explaining the parameters of the case study to students, I communicated that they would not be writing formal English. They would be tweeting, and the two—English and tweeting—were different. Because of my explanation and my qualification, students understood that tweeting was not formal English, and, therefore, they were not expected to use formal grammar and punctuation.

I further explained that communicating with a microblog could have merit if it were used purposefully. This meant that students needed to think for a condensed format: compact expression for compact space and that the limit of 140 characters for writing and communicating would provide a positive structure, not a negative impediment. By putting microblogging in this context, the medium worked. Students communicated purposeful messages, as you will see later.

Whether or not you are a member of Twitter, it is a fact that the social media website continues to gain momentum as a socially acceptable way to communicate ideas, thoughts, feelings, and information to a global audience, in an unfiltered forum that can be democratic, chaotic, or both. I will not analyze the merits of Twitter and other social media platforms that provide users with a quick, widespread means of communication. I will, however, provide an overview of Twitter in classroom instruction.

I conducted a case study with my undergraduate course, Digital Core, the same course described earlier in this book. Again, this freshman-level, introductory course is required of incoming students in the electronic media program in which I teach. The course presents students with an overview of new media methodologies and practices, with an emphasis on electronic media production. So, it was fitting that I would integrate the use of Twitter in instructional modules. The modules focused on microblogging's impact on in-class communication. Students were required to use Twitter and analyze its strengths and weaknesses, in terms of how it facilitated their individual participation in class, as well as the overall experience of microblogging as a communication tool in an educational environment. Some of the results of this case study were unexpected.

Case Study No. 3: Engaging Students with Twitter

The study began with making sure all students had Twitter accounts. As streamlined as the process of creating an account is and as ubiquitous as

Twitter has become, I learned that a large number of my students, approximately 65%, did not have a Twitter account. This was the first surprise. The process of creating an account and becoming a member of Twitter is straightforward. The basic steps are as follows:

1) Create a unique user name preceded by "@" and create a password.
2) Post, or tweet, statements on the website that do not exceed 140 characters in length. Though the length of individual tweets is set, the content of these statements can be anything you want.
3) Elect to follow other Twitter members or not. If you follow them, you can aggregate their comments online for easy access and reading. This became an integral part of the exercise.

I had three objectives in integrating Twitter into the classroom experience:

1) Twitter would provide a new way to engage students with the subject matter via a third party tool. Students who might be diffident or not accustomed to reacting verbally in class would have a new opportunity to be heard. Instead of raising hands to offer answers to questions, students would post their responses on Twitter in class in real time via a laptop with a wireless Internet connection or mobile device. This recalled my use of text messaging in the new media research exercise described earlier in the chapter "Interactive Learning."
2) Because each posting on Twitter is limited to 140 characters, students would be challenged to express themselves succinctly. They would need to distill a response to its essence, a useful exercise in quick, direct communication. Formal punctuation has unfortunately become a liability with the integration of devices and new media into instruction, and it was evident to me that microblogging in the classroom, with its space and time constraints, needed to focus on *what* was being communicated more than *how* it was communicated.
3) Twitter would allow me to aggregate students' responses, that is, tweets, and make them available online. This way all students could read all responses both in and out of the classroom, providing students with the opportunity to further reflect on the subject matter. In

this instance, Twitter would become an archive of commentary in the classroom.

The topic for the class discussion was *global news right now*. Students were instructed to log onto the BBC World News website (http://www.bbc.co.uk), read an article on the homepage, which would hyperlink to a continuation of the article inside the website, and then draft a synopsis of the news story in 140 characters or less. They each sent their relevant tweet to me at a Twitter name (account) that I had created expressly for the class, similar to my creating a unique Facebook account for "Case Study No. 2: Interacting with Literature on Facebook." I intentionally created a brand new Twitter account for this exercise, in order to give the case study necessary focus. This way, it was easy for students to view the tweets populating the interface as they submitted them, because the account they were tweeting to started blank. There would be no previous posts in view that could muddy the results.

After my students posted their tweets, I projected them on a large screen in front of the classroom, so that all students could read the tweeted responses and see how differently students chose to respond. All tweets, provided in the following examples, were copied exactly as they appeared online on Twitter, in terms of spelling, capitalization, and punctuation. Some students successfully summarized a news story, as in the first two examples: one about a change in leadership in Greece and another about the kidnapping of a professional baseball player.

- Greece is in a state of turmoil and there is also a leader shift; Mr Papademos will replace Greece's Prime Minister Papandreou
- a catcher for the Washington Nationals was kidnapped in Venezuela

Other students used the opportunity to provide their own creative spin on a news story. Following are tweets in which students combined their descriptions of news items with editorial comment.

- Looks like we can say goodbye to Rick Perry in the next election
- Sports and Violence: I am finding a awful trend.

Still others ignored the requirement to summarize a bona fide news story and supplied only commentary regarding their in-class experience with Twitter.

- Swimming on a sea of faces ...
- I wonder what twitter would taste like with cheese
- We are all atwitter about Twitter!
- I feel Twitter takes away from personal interaction and there is too much going on.
- Actually NOT wasting time on twitter.
- This is a tweet.
- currently I am tweeting you a tweet that is going to populate into the interface of twitter. I still have 37 characters left oh wai

  (The "t" at the end of the last tweet in the list is purposefully omitted, as it would have been the 141st character, and, therefore did not appear in the original tweet.)

At the end of the tutorials and exercises, all three pedagogic objectives were realized, some with more success than others.

1) The first objective—to engage students via the third party tool Twitter—did indeed encourage diffident students to participate in the discussions with positive results. Microblogging with Twitter gave every student an equal opportunity to contribute to the in-class discussions. Though every student had a voice, certain students expressed themselves more substantively than others. Typically, students who had had experience with Twitter before this exercise were more comfortable using the online platform than those who had none. Because there was no learning curve for students who had Twitter experience, they were more adept at drafting responses that genuinely added to the class discussion at hand. Their contributions elevated the discussion. Students with little or no Twitter experience communicated in a clumsy, less articulate manner. Their contributions tended to be less productive.

   In addition, a majority of students did not follow the specific instruction to summarize a news story. Instead, they tweeted editorial comment or opinion. The examples of these tweets are in the final grouping mentioned earlier. For these students, it was as if the sole purpose of Twitter was to communicate opinion, not fact, to the world. Not even for a moment, not even as a class requirement, would Twitter be a serious communication tool for these students. This realization was telling, and a bit unsettling. To me it indicated that if educators were to move forward with serious integration of microblogs like Twitter into pedagogy, they first would need to assess how students perceive the value of these platforms in education. If students perceive these platforms as having little or no value in education, because to them the platforms are more for goofing around online than for learning, there would need to be a refocusing of students' intentions for using such platforms. This would need to occur before using microblogs in teaching.

2) The second objective—to limit comments to 140 characters—had a mixed result. Some of the students rose to the challenge of thinking meaningfully, in a condensed way, about what they wanted to communicate. Before tweeting, they thought about their message for some time before posting it. Some students even drafted their message in a text editor, like Microsoft Word, and copied and pasted it into the Twitter interface. Pausing before tweeting gave students the chance to communicate their message of 140 characters with more clarity. Other students had little patience or became frustrated and simply typed until they reached 140 characters, posting incomplete comments or ones that did not fully communicate their message. Still other students—who did not view this as a serious learning exercise—were more concerned with having fun. They typed a minimum number of letters just to post something, anything, and what they posted was usually banal and non-applicable. "This is a tweet" is an example of this. As was expected, most students' punctuation was generally messy or absent. This is evident in a majority of the examples provided earlier.

3) The third objective—to aggregate comments, that is, tweets, for reflection and discussion—was successful in that comments were archived and available online. Yes, the tweets are available, but as the Twitter stream becomes longer and longer, with the addition of more tweets over time, it can be difficult to locate the appropriate string of tweets—that is, the ones that pertain to this exercise. For example,

> this case study was conducted on the 10th of the month, and a week later on the 17th, when I revisited the Twitter stream in class, again projecting it on a large screen for everyone to see, I had to scroll for a very long time to reach the class stream. This was clumsy, not user-friendly, and it happened because, between the 10th and the 17th of the month, the stream had become populated with all my students' tweets and retweets, outside of class. Retweets are the messages that your followers receive from outside your stream and then resend. These retweets appear in your stream. Remember, Twitter aggregates tweets and retweets in a single stream from all the users who are following you, your followers.

My objective for revisiting the Twitter stream was to continue the class discussion of the pros and cons of the Twitter experience. There were some of both to discuss. One real downside to using the tool was the accumulation of new tweets appearing after the exercise ended. These tweets were unrelated to the initial class stream, and their presence made it difficult to locate the original, salient tweets, in a long list of messages. The many unrelated tweets, appended to the initial class stream, had no relevance to the class exercise. They created digital noise—textual and contextual. This digital noise compromised the usefulness of Twitter and, for some students, made using Twitter arduous. As one student commented, "Twitter is overwhelming." Because this comment was from an undergraduate student, it underscores one of the findings of this case study, that the appeal of Twitter is generationally non-specific.

This finding—that age was not a factor in determining a passion for microblogging—became more evident when I asked students to give me a one-word adjective to describe their experience with Twitter, whether or not they had used Twitter previously. Before this exercise I assumed, incorrectly, that 19- to 22-year-old students would view using Twitter favorably. Interestingly, about 65% of the students in this case study found using Twitter was a negative experience. This percentage was similar to those who did not have a Twitter account before this case study began, but it represented different individuals. The two groups of 65% were not the same.

Following is a representative listing of the positive and negative one-word reactions students had toward Twitter.

Positive

Simple
Nifty
Informational
Entertaining
Instant
Streamlined

Negative

Boring
Insufficient
Overwhelming
Indulgent
Paranoid
Consuming

Twitter as an archival tool proved less than ideal for several reasons:

1) There was significant digital noise within the Twitter stream.
2) For the most part, students were unwilling to physically scroll down to previously posted tweets, in order to view tweets applicable to the class discussion.
3) The fact that students would not revisit previously posted tweets, unless encouraged to do so, underscored that Twitter is a tool used more for receiving information instantaneously and communicating spontaneously than for revisiting and rereading archived information.

To counter the negatives listed earlier, in the future when I repeated this exercise in another section of my Digital Core course, I made a point of adding a module that required students to revisit Twitter, reread class commentary—no matter how deep into the stream they needed to go—and post a *concluding tweet*. This brought the experience full circle. In the repeat of the exercise, there was more evidence of pedagogical benefit from the use of Twitter. Requiring students to look (scroll) back, read previously posted tweets, and reflect on their overall experience, was beneficial. The following student's concluding tweet summarizes this:

- I get it. Saw the whole picture. Before, during and after.

In *The Medium Is the Massage,* McLuhan describes a kind of civil war, or conflict, that can be created by environments filled with electronic communication, or as he describes it, "informational media":

> It is a matter of the greatest urgency that our educational institutions realize that we now have civil war among these environments created by media other than the printed word. The classroom is now in a vital struggle for survival with the immensely persuasive "outside" world created by new informational media.[17]

This is a call to action for all educators to avoid the conflict. There are no battles to be won by discounting any new medium or technology, even if it is part of the "persuasive 'outside' world." *All* media—inside or outside, unique or conventional, new or old—are persuasive, even invasive in our lives. Rethink why and how we use media in education, and repurpose media when necessary. Use electronic communication, including microblogs like Twitter, to show students *how* to communicate in a new way. New media can produce new messages. Accepting this is key to the successful adaptation of communication in the classroom in our technology-driven world—a world saturated with "informational media."

To all the readers of this book, I say, embrace the new, and let the old inform how you use the new. It is time for all of us, teachers and students alike, to "get it."

[17] McLuhan, *The Medium Is the Massage*, 100.

9

PRESUMPTION OF CONNECTEDNESS

Web 2.0 technologies permeate the consciousness of students in ways we educators may have never imagined. There is never-ending connectedness with digital communications and an expectation that all of us in education, no matter what our demographic, should want this connectedness. It is widely accepted by the majority of the students I teach that, whether via a main communication stream or a backchannel, they should have access to whatever information they want when they want it, as well as communication with whomever they choose whenever they choose to have it. Their lives in front of computer monitors, on smartphones, with tablet computers, and with whatever other devices may be popular, have provided persistent connections to people, data, merchandise, events, information, commentary, entertainment, news, and the list goes on. Once students have all this connectivity, it is difficult to not have it. It is easy to expect it.

In this book thus far, I have advocated for the integration of Web 2.0 technologies in education, within reasonable parameters and with purpose. Sometimes I believe it is necessary to push beyond what is reasonable in order to keep education fresh, innovative, and progressive. Engaging students with interactive learning and including students in the creation of pedagogy—through their participation with tools during learning, which I described earlier as participatory pedagogy—are things I support and practice in my teaching.

But I wonder if I can support *absolute* expectation of and reliance on digital connectivity, which has made its way, seemingly irreversibly, into education. I ask this: Do we want to be connected all the time? Should we be connected all the time? I think most of us, who want to have balanced interpersonal connections with students and balanced implementation of technology in education, would answer no. Therefore, to allow students a *presumption of connectedness*—whereby access to information and communication via technology is constant, without rules to follow or limits to acknowledge—is unorthodox. It may very well be unacceptable.

There is a lack of structure, discipline, and etiquette in a situation where no limits define appropriate connectedness. How can this be good for learning? In order that curricular processes are followed, teaching is organized, and instructors' time is respected, it is imperative to define when students can communicate with instructors, when instructors are available to provide information pertinent to coursework, and when students can expect responses. The current trend is in the opposite direction. More and more, there is a breaking down of the necessary structure that limits access and communication.

I find examples of this in my own teaching experience. In one instance, a student had a question about a project she was working on over the weekend. At 2 AM on Saturday, the student sent me an e-mail inquiry regarding the project. I was asleep and did not respond. At 2:35 AM, just 35 minutes later, I received another e-mail asking if I had received the previous e-mail. Although it is difficult to interpret context or read between the lines in electronic communication, I could tell the student was implying in her second e-mail that I should have responded already. It seemed that the student figured her e-mail would make its way to me in my sleep, perhaps through some kind of digital osmosis, and that I would respond—despite the fact that I would not check my e-mail for another 30 hours or so, on the Monday morning following the weekend, after the sun had risen. This scenario is not at all far-fetched. It happened. Similar instances have happened to me more than once. More recently, another student showed up at my office door without an appointment and with no advance warning and asked, "Did you get my e-mail?" I asked, "When did you send it?" He replied, "Ten minutes ago."

My initial response to this presumption of connectedness is bewilderment. What rational person would expect a response to an inquiry at such an odd hour as 2 AM or within 10 minutes of sending an e-mail? After all, there is basic communication etiquette, in and out of education, that requires all communicators to have a valid concept of time and a sense of appropriateness. This is normal. This is civilized. This is the way respectful adults communicate. When did this sense of appropriateness get lost? Where does the loss of perspective come from? It is complicated.

The students whom I referenced cannot be held wholly responsible for their illogical expectations—that I would be there in front of my computer and ready to answer e-mail in the wee hours of the morning or that I check my e-mail during the day in less than 10-minute intervals. I am not a sociologist or psychologist, but it seems that there are social and behavioral issues involved. There is a societal standard, a pervasive standard, which supports instantaneous responses to electronic correspondence—these days. Across our digital society, value is given to how quickly someone can turn around a response: the faster, the better; the faster, the more important. Many of us have hit *send* before we have finished composing an outgoing e-mail, checked for typos, or attached the necessary document(s) to it. And, many of us feel the pressure to hit *reply*, even *reply all*, before we finish reading the incoming e-mail to which we are responding. To some degree, we all have become conditioned with this knee-jerk response. It seems none of us think enough, or deeply enough, when we use electronic communication.

Let us look at this knee-jerk conditioning in more detail. The standard by which we define quality electronic communication is not so much the content of the message as it is the response time to the message. The discourse that should be contained within our communication, and the provocation of thought that should result from our communication with other people, become secondary to response time. The message is compromised, because we believe that the speed of our response trumps the content of our response. Generally, this idea of a compromised message, for the sake of a quick response, is supported across disciplines, ideologies, and demographics in education, as well as in the professional world.

But my explanation for this presumption of connectedness (in the wee hours of the morning) is more basic. That student acted impulsively, because connectedness seems to be a given. It is part of many aspects of her everyday life. It is not questioned, and, according to the communication standards of undergraduate students and certain sectors of society, it should not be. Connectedness serves fast, facile communication around the clock. So, go with it. Use it. Do it. E-mail me at 2 AM on Saturday, and I will know that you are thinking about the material for class. I would like for a moment to give my student the benefit of the doubt and say, perhaps at that very early hour she had an intellectual or creative breakthrough and wanted to share it immediately. It was such a significant epiphany that it just could not wait for 9 AM on Monday. All of these are good intentions, which could make the presumption of connectedness a good thing. But for productive education, it is necessary to balance zeal with etiquette in communication, like it is necessary to balance theory and practice in instruction. So, it is not a good thing.

I support students' diligence, even if demonstrated impulsively, but I support more the requirement to communicate purposefully. The fact that a student may not know to refrain from e-mailing a professor at 2 AM and expect a quick response is a conundrum that all instructors who use electronic communication need to address. I do. I include a single sentence on all my syllabi that reads as follows: *Student e-mail correspondence will be replied to during office hours*. With this single sentence, my students presume less connectedness. I should add that I make every attempt to respond to e-mails from students as expeditiously as possible, and typically students receive responses from me well before my next set of office hours. But this one sentence, this rule, sets a standard. It helps students differentiate between appropriateness and inappropriateness with electronic communication, something they may never have known or been taught. Or, maybe they just forgot. The rule provides a reminder that not all communication is urgent. The rule provides structure for communication in education.

Presumption of connectedness affects learning. It has to do with attention span. We seem to have evolved into a species that wants things more quickly today than yesterday, consistently. This may be due to advances in technology, more social connectedness created by new media, or some kind of new psycho-social dependence on *being current* and living and thinking *in the moment*. In his book, *Amusing Ourselves to Death*, author and media theorist Neil Postman saw this trend beginning some years ago, in the mid-1980s. Postman attributed the decrease in attention span to an increase in modern technology in our lives. He talks about technology, specifically television's preemption over printed media, as causing a profound change in education:

> We face the rapid dissolution of an assumption of an education organized around the slow-moving printed word, and the equally rapid emergence of a new education based on the speed-of-light electronic image.[18]

I see this trend continuing today, as new media technology has made its way into education in a dramatic way. It very well may be that modern technology is responsible for the limited attention span that students bring to the classroom. Whether it is or not, the perpetually waning attention span of students has an impact on the whole educational process. Like students have come to expect instantaneous responses to their e-mails, they also expect immediacy in their education: faster lectures, heightened in-class stimuli, and more use of technology.

New media platforms and gizmos make education quickly and easily available. Many institutions already have put mechanisms in place to make education more immediate, such as e-learning systems, e-mail alerts, and mobile websites. It is not uncommon to visit the website of an institution of higher learning and see these mechanisms promoted on the home page: Get the new app; check out our new blog; enroll in our virtual community; watch the live webcast, and so forth. There seems to be a consensus among administration, faculty, and students that communication, even that which is official, must be instantaneous. If it is not instantaneous, it is not worthwhile.

Much of the educational process today is focused on immediate communication and immediate connection, and in some situations, immediate knowledge sharing. With all this immediacy, there is little room for gestation, ponderous thought, or deep creation, because students, quite simply, have little patience for process. It is common for educators to stress the importance of process as much as product: The way a student creates a product is as important as the product itself. The distinction between product and process is explored in "Case Study No. 1: New Media Process and Product" in this book. However, in higher education, when students are entering their first term with multiple electronic devices in hand, and the sense that multitasking is a feasible means to achieving productivity (in students' minds), educators must compete with students' device overload and their loud and persistent call for immediacy. In order to accommodate students' expectation

---

[18] Neil Postman, *Amusing Ourselves to Death: Public Discourse in the Age of Show Business* (Penguin Books, New York, 1985), 145.

of immediacy, education must become, and in many instances has become, a results-oriented process.

Educators' process is different than students' process. So, the challenge for instructors who want productive classrooms is to teach or convince students that focusing on one thing is a better way to learn. To do this, educators must genuinely engage students. As mentioned earlier in this book, interactive learning and participatory pedagogy are two ways to involve students in and with their own education, with measureable results. But now, we are talking about working with a kind of technology-induced attention deficit, which can compromise two-way communication and learning. To meaningfully engage students who are preoccupied with wanting results fast, education needs to be oriented with tasks. And these tasks need to be synergistic with learning outcomes.

I am not advocating for busy work. I am advocating for an approach to education with new media that maintains a theory/practice balance *and* engages students with meaningful tasks. One way that I do this is to organize instruction so that tasks are integral to learning: a results-oriented approach that introduces tasks within a balanced theory/practice regiment. It is a multiphase process that I call simply, *results-oriented instruction*.

Results-Oriented Instruction

Phase No. 1: *Talk, Show, and Do* Model

In the courses where I teach first- and second-year undergraduates, I rely on a *talk, show, and do* model. In the first phase, the focus of instruction is more on moment-by-moment *doing*, as opposed to contemplating, analyzing, and formulating. The emphasis is more on practice and less on theory, intentionally.

I create a series of in-class exercises in which I teach tools, while my students use the tools. It is a simple premise: Students learn by *doing*. For example, when teaching jQuery, a framework for creating interactive components and widgets for webpages, I talk about the purpose and function of jQuery first. Then, I demonstrate it. While I type the code to create the desired interactive component, I require students to type along with me—key by key, letter by letter, symbol by symbol. I talk. I show. They do. This *talk, show, and do* model engages students in a more comprehensive way than just *talk and show* might. They perform a task, witness its results, and, by doing what I am showing, they learn.

This model of instruction also motivates students to learn more, because the learning does not seem theory-heavy to them. The theory has been slipped in while they were *doing*. This is similar to when parents disguise vegetables that a child does not like in a food their child likes, for example peas in lasagna. The child eats the lasagna and gets nutrition from the peas without really tasting them. Nutrition gained. Lesson learned! But like the child who will have to eat more vegetables in the future to sustain good health, a student will need repeated exposure to the lesson, so that the topics in it are learned for the long term, and the doing of tasks is not trivial.

If *doing* were the sole focus of instruction, education of new media disciplines at universities might be construed as too in line with the teaching objectives of trade schools. (We want to avoid this, as mentioned earlier in this book.) So, how does an instructor strike the necessary theory/practice or *why/how* balance in teaching, when students tune out as soon as tasks end and theorizing begins, and, more importantly how does an instructor sustain students' knowledge of topics for the long term? The answers lie in the second and third phases.

Phase No. 2: Academic Conditioned Reflex

In the second phase, think Pavlov. Most of us are familiar with the Russian physiologist Ivan Pavlov, who, in the late 1890s and early 1900s, conducted experiments with dogs, in which he caused them to salivate by ringing a bell. Through a process that came to be known as *conditioned reflex*, Pavlov sounded the bell every time he fed a dog. After repeated association—bell/food, bell/food, bell/food, bell/food—the dog would salivate when the bell was rung and no food was delivered. Working with new media in education, we can create a similar kind of reflex, with the result being students grasping theory by associating it with repetitive practice or tasks.

Using the jQuery example mentioned previously, I have my students repetitively write the code to create various iterations of a single widget. Each iteration is similar enough that writing and rewriting the code constitutes a repetitive action but different enough that the learning process is organic and results in a different product each time. The specific task is to create a slideshow using jQuery for a webpage. In the first iteration of the slideshow, students are required to present images. In the second, they alter the code to present various text passages. Then, in a third iteration, they alter the code to change the direction of the slides from horizontal to vertical. And in a final, fourth iteration, they alter the dimensions (width and height) of the slideshow, so that the widget can be used in a different webpage layout.

Into this whole process, I incorporate lectures on interaction design and theories of HCI, so students understand that the primary objective of writing code and integrating a widget into a webpage is to engage the user in a meaningful screen-based experience, not just to create a digital effect *on click.* Having worked with basically the same code *repeatedly*, students can write it. And, having heard me lecture *repeatedly* on the purpose of the widget, students understand the theoretical underpinnings of jQuery, as well as the scripting language at the core of jQuery, JavaScript. My objective is to tie the two together as an *academic conditioned reflex*. Code/lecture, code/lecture, code/lecture—students learn it. The long-term pedagogic goal is that in the future, when the students write similar code—having learned the theories of interaction design along with programming code—they will create meaningful user experiences with their code. And they will be able to adjust the code as needed.

Phase No. 3: Solidifying Behavior

In the third phase, the goal is to solidify the task-oriented behavior created in the previous phases and make it last. In this phase, the tasks that were the foundation of pedagogy in the second phase, and the learning that resulted from the repetitive *doing*, are transformed into lasting behavior—which will benefit students moving forward in a new-media-focused curriculum. Task-oriented conditioning has long been a part of teaching new skills, namely to structure students' behavior. Therefore, using task-oriented conditioning to teach and motivate is not novel. However, using it with solid pedagogic practice, alongside new media technologies, is key to ensuring that tools remain a means to an end, not an end unto themselves.

Continuing with the jQuery example used previously, in order to reinforce the usefulness of JavaScript in the design and production of online properties, and at the same time solidify task-oriented behavior in students, I expanded instruction to include new applications of JavaScript. Students explored integrating JavaScript into media-rich webpages that would give users control over media assets, both video and audio clips, in the form of media players. This was similar to the exercise of creating media players in the chapter "Backchannels and Multitasking," but this exercise was more complex. In this exercise, students were required to apply the knowledge they had gained about JavaScript and jQuery in previous lessons to a set of new objectives. They needed to apply cumulative knowledge. Therefore, they learned more advanced concepts surrounding jQuery, and equally importantly, they solidified what they had previously known.

Making the conditioned behavior last is beneficial to students in new media disciplines, because it gives them the flexibility they need to work with a varied and evolving set of tools and technologies, and it empowers them to learn how to figure something out when they do not readily have the answer. If students can produce fully realized work with a current tool, technology, or scripting language, they can easily apply their behavior and their newly acquired body of knowledge to new tasks, with new challenges, and with new tools, technologies, and scripting languages in the future. Behavior, learned through repetitive doing, is transferred to new tools with new objectives for producing digital works. More importantly, task-oriented conditioning prepares students to *learn how to learn*. That is key with new media technologies.

To recap, the three phases of *results-oriented instruction* are as follows:

Phase No. 1: *Talk, Show, and Do* Model
Phase No. 2: Academic Conditioned Reflex
Phase No. 3: Solidifying Behavior

These three phases provide a structured approach to teaching students. Though initially focused on doing, the model ultimately supports the creation of curricula that balance doing with thinking. Certainly, this approach is beneficial to teaching disciplines related to new media, but it is also a structure that can easily be applied to teaching other disciplines. Because new media are relevant to many other disciplines, I am often asked to guest lecture in other majors, including business, journalism, design, arts administration, engineering, and even physics. Though I lecture on new media itself, my approach to teaching, which could be described as *thinking while doing*, is frequently adopted by the professors whose classrooms I visit. I have come to learn that a results-oriented approach to instruction works across disciplines.

Having witnessed the success of results-oriented instruction with their own students, some of my colleagues add modules to their courses that incorporate this approach to teach topics as varied as e-business, magazine writing, community engagement, and music. For example, in an electronic advertising and marketing course, the professor saw that having students produce a viral video for an online marketing campaign—instead of just talking about it and presenting the concept in a rudimentary slide presentation—made sense and illustrated how students benefited from a *thinking while doing* classroom experience. Students' marketing campaigns were more informative and more compelling than those produced by a class during a previous term, when a results-oriented strategy had not been implemented and there had been no requirement for students to produce their concepts beyond discussion and verbal presentation. Just thinking did not produce better results. Students had to think *and* do.

In another example, arts administration students were required to conceive of a website for an arts organization that would increase membership and patronage. Students had solid ideas, but until the ideas were tested in a live website—which they produced with basic web design and production tools, as they had little hands-on experience—students were not able to present a definite communications strategy for the organization. They created a more focused strategy for the organization and made better decisions about how to implement that strategy with supplemental marketing platforms when they built the websites themselves. In building the website, they understood the nuts and bolts, graphics, and code supporting their strategy, and they were better informed about how to integrate interactive and social media components into the website to support their strategy and, hence, market their organization online.

Because devices in some form or another are here to stay, I predict that students' presumption of connectedness will not go away any time soon. Results-oriented instruction is a way to work with the presumption of connectedness in order to provide students with a regiment for solving academic- and technology-related challenges in day-to-day learning. Results-oriented education also empowers students to discover new ways to apply knowledge, which they have previously acquired, to future projects through behavioral conditioning. Teaching students, in and out of new media, with the three-phase structure described earlier, prepares students to be thinkers who do, as well as doers who think. This is necessary for success in these digitally transformative times in education.

10

INTERACTIVE CONTENT AND ONLINE AGENDA

The infinite repository of interconnected, digital content on the Internet that lies behind computer screens is alluring. Any demographic of user can easily engage with these worlds in a process that has become so very familiar: log on to the web, access one bit of information, then click to find more, then scroll to see more, then swipe and click for even more. Going back to the time when the Internet first became a publicly accessed medium, it was not unusual for users to click on hyperlinks for hours on end. One leads to another, then to another, then another. This practice of non-linear surfing, or becoming lost in webpages, continues today. It is a practice that, because of the intuitive nature of surfing the web, immerses the user in online content through a process of discovery.

This activity is one of the reasons for the success of wikis, such as the popular Wikipedia. It is a captivating journey, when one immerses oneself in the hyperlinked content of wikis, where a curious tour can start on a webpage about New York City and after some clicks end up on a page about the Synod of Dort: a meeting in 1619 Holland to settle a controversy initiated by the rise of Arminianism. This is the kind of journey that can only happen in the seemingly endless web of online content.

Beyond being expansive, wikis are powerful communication tools. In theory, wikis provide for the democratic and unfiltered contributions of content to online platforms, which can be visited by web surfers worldwide. Wikis are websites that, with their server-based technology, allow users to create and edit content and view the changes they made to that content live in a web browser, usually on the fly and in real time. Take Wikipedia (http://www.wikipedia.org) for example. This website is an online encyclopedia where users can update information that anyone in the world has posted and amend the postings as they wish, ostensibly. Wikipedia, like other wikis, also serves as a repository of information that, good or bad, valuable or not, has become a resource for both curiosity seekers and researchers. In theory, wikis provide interactive environments for the free exchange of ideas and knowledge, allowing the whole world to participate in the creation of content online.

I have described the mission of wikis with this condition: in theory. This is important, because free-flowing expression and fact do not necessarily mix. In practice, in order to be reliable, wikis must filter their information, at least to a degree. Wikis that have been online for a considerable amount of time have repeat users. Loyal users would not and should not tolerate erroneous information. So, in order to have longevity and be construed as a trusted resource, a wiki must monitor postings and filter them. The filtering happens behind the scenes, often unbeknownst to the user, by a staff of people employed by the wiki, whom I will call *wiki elves*. For example, edit a post on Wikipedia, and you will see that the change does not go live immediately. It is being reviewed. And, if there is objectionable content added, *objectionable* as defined by Wikipedia's management, the content may never go live.

There is some level of deception here. The freedom of online expression—change it and it will appear—is not total. With filtering, information sharing that, in theory, is integral to the wiki experience, is marginalized. More than that, some information in wikis is plain wrong. Wikis that are controlled by parties focused on communicating *their* message and putting forth *their* agenda can easily do so with filtering. These kinds of messages, which may likely contain misinformation, could be harmful to a trusting global public.

So, how do you know if the information is real? How can you trust what you read? What about the democratic exchange of ideas? All of these questions about wikis lead to an even more important one: Are wikis appropriate in education as a tool for teaching? I set out to answer that question with this next case study.

Case Study No. 4: Analysis of Content in Wikis

In this case study, I integrated the use of four kinds of wikis into a class module on the analysis of online content. Students were tasked with determining if the content presented in each of the four wikis was of real value or if it was misinformation, perhaps agenda-driven. They would do so by answering the following two questions:

1) Does the wiki provide useful, unbiased information that is consistent with the mission of the wiki?
2) Can the user contribute content to the wiki that will add to the wiki's overall usefulness?

The four types of wikis employed were as follows:

1) An informational resource wiki, like Wikipedia.
2) A communications wiki, specifically one focused on the broadcast industry.
3) An arts wiki.
4) An education wiki, specifically one focused on higher education.

Students visited each of the four wikis. They were instructed to contribute content to each wiki that would make it better, richer. What students found upon logging in to the wikis was interesting and varied.

Students found that with behemoth wikis that were well-established and well-funded, like Wikipedia, new content would be reviewed before going live (as explained earlier). Here, democratic expression and immediacy were compromised.

Students found dated wikis that were created for specific past events, like education conferences. Some of these had interesting postings, especially if users were interested in the niche topic of the conference, but the content was years old, and the wikis were closed—no further content edits or additions could be made. Here, obsolescence was an issue.

Still other students came across wikis that were industry generated. Commercial groups have moved into the wiki world, in an effort to promote their clients and products to the web-surfing public. These groups are focused on using wikis for public relations, marketing, and advertising purposes. Here, there was complete message control by the groups hosting the wiki. For example, one broadcast company was promoting a new television series in a wiki. When students edited information about the characters in the series, the students' changes appeared live, but only for a brief period of time. When students logged on the next day to check on the status of their changes, they found that the broadcast company had revised the content back to the original, which supported their marketing agenda. It was evident there was no true freedom of expression, and, from feedback I received from my students, there was a sense that the user was being duped.

This was a useful learning module for students. Some students who, prior to participating in this exercise, had a sense that the Internet would only provide the best possible information to the public, learned otherwise. They saw that agenda can be a part of their online experiences and that digital democracy can sometimes be digital autocracy. Prior to the exercise, some students did not know they could participate in altering content in wikis—a fundamental purpose of wikis. Though most of them had used Wikipedia, some students assumed it was a non-interactive online platform, a website that merely presented information.

By learning that they could contribute their voice to a global platform by changing hypertext and adding hyperlinks, students acquired a greater appreciation of the interactive nature of wikis. In some instances, students felt empowered—even though they would learn that their voice would be silenced, their contribution deleted, when the behind-the-scenes wiki elves reverted text back to its original form.

Most importantly, the exercise provided students with the opportunity to think critically about the presentation of content online, to distinguish between a filtered and unfiltered environment, and to take the time to analyze content in order to assess whether it has merit.

At the conclusion of "Case Study No. 4: Analysis of Content in Wikis," I was not able to answer all the questions I set out to answer. So, I decided I would explore wikis in a little more depth for myself. I wanted to see, firsthand, if the automatic control of content by wiki elves is becoming more and more an issue, as I suspected it was. I logged onto the popular Wikipedia. This "mother" of all wikis would be my testing ground to see whether or not online democracy, in terms of making contributions to a content stream, was genuine.

I visited a Wikipedia page about a published literary work that has a relatively wide international following. I read the posting and noticed that some of the information, which had been posted by an anonymous user, as most are, significantly misrepresented the work. Considering the blatant misinformation in the posting, I assumed it had been posted without the permission of the owner of the work, who, as the rightful owner of the intellectual property, would have an interest in the work being properly represented, especially on such an expansive platform, or so it would seem.

I had not created this Wikipedia posting. Someone else had. Actually a community had—community as defined (or better, *re*defined) by Wikipedia. But as an academic with an appreciation for authenticity, a mandate to present factual information with accuracy, and a respect for intellectual property and those who create it, I felt something should be done—a wrong should be made right. Because nobody in the Wikipedia community seemed to have done the proper research to verify all the necessary facts and present the work accurately and without bias, I decided I would do the research, in depth and impartial, and contribute information to the Wikipedia posting that would correct its inaccuracies. After all, this is a wiki, a forum for democratically aggregated content. We are all contributors. We are all members of the community. Wrong.

I attempted to alter the online text in the Wikipedia Edit section. But I was not able to. After typing my changes, I pressed the *Save Page* button, and I received a computer-generated message alerting me that changes could not be made to the content in the posting. In this particular experience, the behind-the-scenes wiki elves were *bots*—computer programs or software applications that run automatic tasks on the web—and the bots, not the users or members of the community, controlled the Wikipedia content. The bots would not allow me to make necessary changes. The bots had been put in place by the community, whom I would come to learn did not *want* changes made to the posting. They had an agenda, which was at odds with the mission of Wikipedia. I felt duped. This kind of automated control is unfortunate, and it speaks volumes about the perceived, inconsistent, and sometimes non-existent democracy in wiki communities.

I was not able to alter the content that misrepresented the literary work, but still I wanted to right this wiki wrong, so I sent an e-mail to Wikipedia—to the *info* address, which is the only e-mail address readily available on the website. In the e-mail, I requested that the content on Wikipedia regarding the work, which, according to my research, was being misrepresented by the

Wikipedia community, be changed and that the erroneous information be corrected. I explained that because of their *bot*, which prevented changes from being made by anyone outside of the *re*defined community associated with that particular posting, I was not able to change the posting. Here is part of what I wrote in my e-mail to Wikipedia:

> I attempted to edit the [Wikipedia] article, but the changes were reverted back to the original text, and I received a message stating the [bot] had restored the original text.

Wikipedia's response to me read as follows:

> We understand that the [work] per se is someone's intellectual property…

Per se! The Wikipedia representative implied that ownership of the intellectual property was conditional. *Per se*! And then the representative continued to defend Wikipedia's behind-the-scene practices.

Authorship is absolute. Ownership of this particular work is absolute. The work is published by one of the largest and most reputable publishers in the world. It is registered with the Library of Congress and protected by U.S. copyright law. No new media platform has the legal or moral right to marginalize ownership. Intellectual property cannot and must not be fodder for members of communities, who seek to tweak, edit, and control content, without respecting published works or the authors who write them. No wiki—even one that is collaboratively and freely edited by a global community of users—has license to alter content and distort context so as to misrepresent other people's works. I fear that Wikipedia and other wikis may be treading into areas way beyond democratic expression and freedom of speech. This concerns me. There is something unsettling about content creators being sidelined by technology, when these creators simply want to protect their intellectual property and the integrity of their work.

So, after conducting "Case Study No. 4: Analysis of Content in Wikis" and interacting with Wikipedia, I conclude that in practice, wikis are flawed. Like a number of the new media technologies and platforms examined in this book, wikis wield too much power and influence and are not interested enough in balancing authenticity with user input, fact with opinion. However, that being said, in terms of providing students in the case study with an interactive forum for the exchange of ideas and a platform to democratically engage with online content, there is something exciting about the concept of a wiki. In theory, wikis are powerful.

11

THE COST OF TECHNOLOGY

I am a technology guy, in terms of understanding new media technologies and integrating them into my teaching. I get the appeal of gizmos and gadgets. I understand the power of some technologies and the silliness of others. But I am a teacher first. I want to be sure that what I use to teach students, empower them, and open their minds, will serve their education and future careers. If, in my pedagogy, I rely on some technology just *because* it is mobile, wireless, smaller, and faster, I have simply taught my students how to rely on something that is mobile, wireless, smaller, and faster. That would be a disservice to my students, because tomorrow there will be some new technology that is *more* mobile, wireless, smaller, and faster. And the substance of what I teach my students, I hope, is greater than inevitable obsolescence. I want what I teach my students in their device-driven, technology-focused education to be lasting, to go on.

In a recent book about using mobile devices in higher education, published by a well-known academic publisher, the book's author asserted that higher education is missing the boat when it comes to fully integrating new technologies, namely mobile devices, into teaching protocols. The author cites examples of how other arenas, such as the corporate sector, agriculture and farming, even retail advertising and marketing, are using mobile devices to communicate, conduct business, and reach clientele. The author implies, quite emphatically, that educational institutions are facing a perilous future if they do not board the mobile luxury liner and sail into the future with the latest emerging technologies in teaching. In essence, the author advocates putting a smartphone in every professor's briefcase and a computing tablet in every student's backpack. I could not help but think that this assertion was dramatic, even hyperbolic. It is difficult for me to believe that my professional ship will sink if I do not have a certain mobile gizmo keeping my pedagogy afloat.

I tried to understand where this author was going with his line of thinking. Was *his* ship sinking? Why must we in education feel compelled to integrate technology in our teaching—sometimes absolutely, sometimes at any cost—when a more holistic approach seems wiser? Certainly, we can always add new dimensions to what we teach and how we teach it, and new media technologies can help us do that. But I continue to support a model for education that integrates new media technologies with tried-and-true protocols and older technologies that have been shown to work.

We professional educators do not have to expunge traditional methodologies and conventional tools in order to integrate digital bells or technological whistles into teaching. We can have both. We should have both. It is the balance of the conventional with the new that makes for the successful integration of technology into teaching. We are shortsighted if we think the way we have been teaching for generations is suddenly obsolete to the point of "missing the boat." It seemed that—with the assertions that worthwhile instruction today *must* include the use of mobile devices and that education without these must-have devices is inferior—the author's line of thinking was headed toward a pedagogic iceberg.

The author's rhetoric is informed by what I call *technological elitism*: the view that what is new is good, what is expensive is better, and what is *new and expensive* is best—in terms of technology. The builders of the ill-fated luxury liner *Titanic* shared this general view. We know where *new and expensive* got them. In higher education, such elitist rhetoric works on the fears of students of all ages, who want an education but do not have a plenitude of new and expensive devices. The rhetoric works on the fears of teachers, who may believe that solid instruction is not dependent on the latest technology. It also works on the fears of administrators, who are tasked with keeping their institutions competitive and financially afloat—especially during economically difficult times, when competition for attracting new students and increasing overall numbers of students, or at least keeping the numbers even, is more intense. The author's rhetoric even works on the fears of family members of students, who want the best for their children, grandchildren, siblings, spouses, or cousins, and who might be willing to provide the tools but cannot financially do it.

The author's assertion can unfairly create anxiety. Students, who cannot afford required technology may become anxious or even self-conscious, and this can have an adverse effect on academic performance. Anxiety, resulting from the inability to afford or acquire the hardware and software, is a concern. However, I have written in this book that we need technology in education. In the chapter "Tool Literacy" I made the case for tool literacy in education. We cannot create tool literacy without tools—obviously. Yet, there is the real issue of making tools mandatory, when certain economic or social conditions make it difficult for students to acquire the tools. There is a lot to consider and to resolve.

Beyond finances, disenfranchised students, in and out of established Western nations, may not have technology for a myriad of reasons:

1) There may be government restrictions preventing access to or purchase of certain technologies.
2) In geographically remote places, there may be physical and/or topographical impediments preventing people from acquiring technology.
3) Certain religions may shun the purchase of technology.
4) Acquiring technology may be a matter of choice. Some people, for whatever reason, may simply choose to not purchase technology.

How do we justify requiring it all—the new tools, technologies, and devices—when populations of students come from various socio-economic and cultural groups and there exist sundry legitimate reasons for not purchasing technology? We do so by making the case strongly and convincingly that, in this new media age, technology adds a necessary dimension to students' educational experiences. We must also maintain that students' success in the classroom is dependent on several important things: hands-on experience with tools, insight into theoretical applications of technology, the acquisition of a certain level of tool literacy, and an overall understanding of how devices are integrated into learning.

The reality in education is that many disciplines require technology. Technology is expensive. Supplying every student with tools, technology, and devices expressly for the purpose of an equitable, affordable education could be cost prohibitive. If an institution or school district cannot supply these tools, the financial burden falls on the shoulders of the students and/or the people who financially support the students. In terms of education, this presents a situation that is potentially *un*fair and *in*equitable. Those who want education to be available and accessible to all demographics may even find this unsettling. Students who cannot afford the necessary technology when it is required will likely become disadvantaged in their education and professional livelihood, post-education, without it.

If we do not want to perpetuate a new *digital divide*—the digital divide of the late 1990s and early 2000s was a topic of much discourse, as the Internet quickly became a ubiquitous means of communication—we must find a solution. Perhaps the burden is on institutions to find more resources, in order to make more technology available to more students at an affordable price. Perhaps the government should foot the bill. Perhaps some students may not be able to study certain technology-based disciplines, because technological elitism trumps their need to have the technology, and they cannot afford the tools. Or, perhaps educators need to be creative in how we require the tools and technologies and even more cogent in making the case for integrating them into the educational process.

Despite efforts to make technology more available, the digital divide seems to be growing wider, as persons of financial means are likely to own certain devices—namely smartphones, tablet computers, and the most recent handheld devices—and economically disenfranchised persons are not. This is part of a new socio-economic phenomenon called the *app gap*, named because much of the computing done on these devices is done via applications, or *apps*. What makes the disparity between the have's and have not's especially interesting, with mobile applications in particular, is that the gap is sometimes created when a child is very young. Children, as young as a few months old, are equipped with handheld devices onto which apps are downloaded for play and education. The children of wealthy parents are more likely to have these devices. An October 2011 study conducted by Common Sense Media, a not-for-profit organization focused on informing the public about the trends in technology, media, and education that affect children and families, stated the following:

> More than a third (38%) of lower-income parents don't have any idea what an "app" is, compared to just 3% of higher-income parents. And just 14% of lower-income parents have ever downloaded any apps for their children to use, compared to 47% among the higher-income families. All of these disparities come together to contribute to a large gap in use of new mobile media devices: 22% of children from lower-income families have ever used a cell phone, iPod, iPad, or similar device for

> playing games, watching video or using apps, compared to 55% of children from higher-income families.[19]

At present, according to a study conducted by ECAR (EDUCAUSE Center for Applied Research), 55% of undergraduate college students own a smartphone and 11% own a tablet computer.[20] These numbers do seem unusually large, but they will grow in the future for all income brackets, as the price of devices, mobile and otherwise, falls over time—which is historically what happens. This is encouraging for technology in education. However, technological elitism may prevail once again, as the iterations of devices that students—from different socio-economic groups—have will vary. Affluent students will likely possess the latest versions of devices. Poor students will not. This kind of socio-economic divide becomes even more apparent, and serious, in second and third world countries, where obtaining technology is only possible for the financial elite. Here, educators must work extra hard on the balance between the real and perceived need for tools.

In my years of experience teaching topics related to new media at urban and suburban colleges, at American and European universities, I have learned that if students truly understand the educational benefits technology can provide, they willingly acquire the technology—sometimes with financial assistance from relatives, sometimes by working hard to pay for the tools—because the case is made repeatedly by faculty and administrative leadership for the relevance of technology in students' education and career preparation. For example, I teach a program that requires students to purchase a certain high-end laptop and software suite. Because the equipment and tools are standards in the industry, it is clear that the technology serves a valid purpose. The technology duly prepares students for professional lives in electronic media or careers closely related to it. The purchase of the technology is an investment in their education *and* their future.

If the case for the purchase of technology is not sufficiently made, students may retreat from the educational process. When students retreat from the educational process, because they cannot afford or acquire the *right* technology—the right technology is that which they need for their education and

---

[19] Victoria Rideout with Melissa Saphir, *Zero to Eight: Children's Media Use in America*, (Common Sense Media, San Francisco, October 25, 2011), 21.

[20] Eden Dahlstrom, Tom de Boor, Peter Grunwald, and Martha Vockley, *National Study of Undergraduate Students and Information Technology, 2011,* (EDUCAUSE Center for Applied Research, Boulder, CO, October 2011), 7.

which has been properly implemented into teaching by educators, versus the technology they use for social interaction in everyday life—they could be made to feel inferior academically. Even if there are legitimate reasons for students not meeting the requirement to acquire certain hardware, software, or devices—however these requirements are created or imposed—there can be the belief that students without the technology are not as good as or competitive as those with it. This can create *digital inferiority complex*: a condition where an individual feels inadequate, because of tools and technologies he does not have or cannot possess.

Digital inferiority complex is counter-productive in learning, and it has no place in education. Students should be empowered by the use of new media in the classroom. Some disciplines do not need new media gizmos all of the time. It is in all of our interests—teacher, students, and administrators—to carefully and thoughtfully determine technology requirements, so that tools in education matter, and they are not gratuitous. We should never use technology just because it is there.

When students understand the benefits of technology to pedagogy, there is no undue pressure on them to buy the technology. The technology is not gratuitous, so students buy it and *get it*. They get the importance of it. There is no technological elitism here. There is little risk of digital inferiority complex. Instead, there is a sense of shared responsibility, a kind of technology-acquiring democracy, in which all the stakeholders have a voice: students, instructors, educational administrators, and, where applicable, students' family members and financial supporters. And, because the stakeholders are upfront about cost, there is fairness and transparency. Students are made aware that the amortized cost of their hardware and software over four years—the time they will likely spend in an undergraduate program completing their degree—is minimal compared to the benefit of the investment.

Along with this explanation, we must continue to work with school systems, government organizations, and the manufacturers of hardware, software, and devices to make the technology available and affordable for all students. Some educators may not do enough to keep technological elitism out of education. And, some educators may perpetuate technological elitism, by allowing manufacturers of technology products to influence their teaching agenda. This is a new media danger zone. In a November 2011 article, *The New York Times* reported the following:

> The demand for technology in classrooms has given rise to a slick and fast-growing sales force. Makers of computers and other gear vigorously court educators as they vie for billions of dollars in school financing .... Apple invites educators from around the country to "executive briefings," which participants describe as equal parts conversation, seminar and backstage pass .... Some critics say the trips could cast doubt on the impartiality of the officials' buying decisions, which shape the way millions of students learn.[21]

I teach with Apple devices and tools. They are among the many different kinds of new media technologies and platforms I use. Over time, I have determined how to best integrate tools in my pedagogy, no matter who the manufacturer, in order to teach my students the theories and skills I believe they need to be successful graduates. Therefore, it is the tool itself that matters, and what can be done with it to create meaningful digital work, not the corporation that produces the tool. If I were one of the teachers invited to Apple headquarters to take a tour of the facility, have lunch with executives, and listen to pitches about new products, I might be persuaded by the technological wonderland I witness there enough to rethink my teaching agenda and the technologies I use to meet learning outcomes. I might be influenced by the impressive buildings, fancy devices, bright smiles, and tasty food. I might become a techno-elitist, whose commitment to educate students democratically and without economic prejudice would be secondary to the *wow* factor. I just might.

If all this happened, I would be headed toward the same pedagogic iceberg as the author whom I referenced earlier, who believes that education is incomplete without certain devices. I would make sure, after my trip to Apple's executive offices, that I was not persuaded by technological razzle-dazzle, unless it was good for my teaching. The tools that Apple or any manufacturer produce must support learning outcomes. This must remain a priority. We must look beyond the sales pitches given to us by software salesmen and hardware vendors, because there are no quick fixes, especially at high prices.

If we want to integrate new media into our pedagogy successfully and justify the cost of technology, we need to carefully analyze and test it first.

---

[21] Matt Richtel, "Silicon Valley Wows Educators, and Woos Them," published November 4, 2011, http://www.nytimes.com/2011/11/05/technology/apple-woos-educators-with-trips-to-silicon-valley.html?hp

When new media is not fully tested before being integrated into curriculum, the danger zone takes on another dimension. A review of some of the learning systems presented in *The New York Times* online show how some widely used systems fail to meet expectations and get results:

> A federal look at ten major software products for teaching ... found that nine of them, "did not have statistically significant effects on test scores." Amid a classroom-based software boom estimated at $2.2 billion a year, debate continues to rage over the effectiveness of technology on learning and how best to measure it.[22]

The bricks and mortar of higher education and all the people and things we put inside the bricks and mortar cost money—especially technology. Technology is often used as a benefit when promoting a school. Colleges and universities that can boast of state-of-the-art labs, hybrid production studios, experimental teaching facilities, digi-toriums, immersive caves, and a cavalcade of other technology-focused spaces, are likely to impress prospective students when showcasing these spaces during recruiting events. This goes back to the notion that *new and expensive* is best. Is it fair to use technology in this way, by implying that education with technology is better ... best? Let us take the following example:

> University A has been ranked by a popular news magazine as one of the top universities nationally for integrating mobile technology into curricula campus-wide. University A is requiring its students to have smartphones, computing tablets, and a collection of applications and software. It follows, then, that other universities will consider requiring students to have certain devices, as a necessary means to getting a top ranking. This is higher education's version of keeping up with the Joneses.

This kind of competition is real. These kinds of rankings, for better or for worse, are important in the marketing of higher education. I taught at a university that prided itself on being one of the nations "most wired campuses." The university pitched this status in its marketing and public relations cam-

---

[22] Trip Gabriel, Matt Richtel, "Inflating the Software Report Card," published October 8, 2011, http://www.nytimes.com/2011/10/09/technology/a-classroom-software-boom-but-mixed-results-despite-the-hype.html?_r=1

paigns. Of course, I understand the business of education and the need to sell one's material assets, and there certainly are instances where this kind of promotion is justified. However, we need to put the marketing of technology in education in perspective, and perhaps rethink it. We can sell our new media assets without asserting that the assets are more valuable than the education itself.

We also need to reinforce the fact that learning is independent of technology. Really. That statement may come as a surprise from this author, a self-proclaimed technology guy. But the expectation that a freshman entering a four-year degree program should have a lot of technology *in order to learn* is a false one. We are setting up students to assume they cannot learn without the technology. Students will not learn just by *having* technology—not through osmosis, wishful thinking, or substantial credit card debt. I believe the technology, tools, and devices we integrate into pedagogy help students to learn, and I assert that education in a discipline like electronic media is incomplete without hands-on experience with technology. But the technology itself is not necessary to learn, because technology does not teach. Teachers teach.

Those educators who think differently, and who would impose substantial, unnecessary technology requirements on students, are taking the digital divide to a whole new level. This is a concern. Just as there must be a balance between theory and practice in instruction, there must be a balance between the real and perceived need for tools in education. We need to answer the following question: What technologies do we *really* need to create tool literacy in future scholars and professionals? The answer to the question is simple: the technologies that *really* serve pedagogy. Only instructors and curriculum developers can determine that.

There are some technologies that bring meaning to teaching, and then they fall out of favor. They lose their effectiveness, or *wow* factor, or both. There are many reasons why a technology falls out of favor. In the marketplace, this can happen when a technology loses market share or is bought by a competitor and then discontinued. In education, this usually happens after a technology has been tried and tested in real-life classroom situations and ultimately it fails to support learning outcomes. When this happens, an instructor would discontinue using the technology, rightfully so.

It is easy for the most pragmatic educators to choose technology because they can, not because they should. As we devise curriculum that is dependent on handheld devices, tablet computers, and other new technologies, let us not participate in the *technology as status* culture, and let us not get caught up with the *wow* factor. Let us work to mitigate technological elitism and eliminate the digital inferiority complex. Let us choose technology that will best serve learning outcomes and the core values of education—at a cost that is reasonable. Then, new media in education will be transformational. And we will avoid the iceberg.

12

MOBILE EDUCATION

With mobility, learning does not have to be site or time specific. Instructors can teach 24 hours a day, 7 days a week, and students can learn wherever and whenever they want—at two o'clock in the afternoon or two o'clock in the morning. They just have to turn on their tablet computers, download course assignments, and learn. But what kind of learning is it? It is not conventional learning in the age of chalkboards, where students had to physically be present in the classroom and take notes with paper and pen, before the wipe of an erasure cleaned the notes away. But it is conventional learning in the age of technology.

It is accepted in higher education that e-learning systems, online platforms, and handheld devices are conduits between teacher and student and that information delivered by teachers and received by students on these systems, platforms, and devices will move fast and be available easily. It can be, because the information is electronic. It is digital. It travels around networks at impressive speeds, and it can be stored on computers for very long periods of time. It is malleable and transferable. Because of this, there is an expectation that digital content accessed on mobile devices is good for education.

One objective of educating with technology is to provide students with the tools they need to learn in a digital environment that is screen-based and interactive, as well as the knowledge and wisdom to use the tools effectively. Another objective is to provide students with the experience of learning in a screen-based environment, and the understanding of how it is different than learning in a traditional classroom setting, but still equally valuable. Achieving the latter objective is one of the biggest challenges with mobile education—made possible by content that is potentially available all the time—because it is education that lacks a certain, and necessary, structure.

Effective learning results from structured teaching. In order for teaching to be effective, structure must be specific to learning goals and established in road maps that guide students to meet these goals. In many instances, these road maps are course syllabi. Instructors draft syllabi with specific learning outcomes in order to establish a set of objectives they intend to meet in instruction and to create the pedagogy necessary for students to meet these objectives. The educational trajectory for students—that is, their process of acquiring knowledge by progressing through course modules or teaching components presented in a syllabus—does not have to be linear, but it does have to be organized and set.

The digital environment of mobile learning does not fully support this kind of organization. Because a mobile learning environment is more free-form and less structured—with its ready-to-serve promise and 24/7 potential—than a brick-and-mortar learning environment, pedagogic structure is compromised. One of the main reasons the structure of education is compromised is what I describe as the *willy-nilly factor of mobile education*: Content is out of context and available around the clock.

When teaching materials—such as syllabi, tutorials, and lesson plans—are in digital file formats and stored on network servers, the materials can linger in cyberspace for an indefinite amount of time. When this happens, the educational trajectory can become indirect, even vague or directionless, because students can access learning materials at various times, from various locations, and on various devices. If students access information willy-nilly—when they want it and from where they choose to log on—information is not necessarily communicated or delivered in the proper context for learning. Course material may not be presented coherently.

> For example, students could access the module for a particular class in the middle of the night, in front of a television broadcasting reruns of a 1950s sitcom, or on a laptop, where a music video from the latest rap star plays side by side in another browser window. Neither the sitcom nor the rap video has any relevance to the learning module. Both are potential distractions. Mixing subject matter with distraction muddles learning, even for the student who claims to be a masterful multitasker. And as we

know from the chapter "Backchannels and Multitasking," multitasking is rarely favorable to productive learning.

When education is presented in an e-learning system, distance learning module, or other technology-based platform that allows the student to be in a remote location, it is likely that the information will be communicated in a way and a time that students want to receive it, not in a way and at a time when the instructor wants them to receive it. This haphazard access to instructional data, when education is mobile, can lead to ineffective learning with new media. This can be avoided by establishing specific times when course materials will be released to students. Blackboard, a popular e-learning system, as well as other systems used in distance learning, have a control mechanism that allows instructors to determine when files and information posted on the system will be available to students, by both date and time. These systems also have built-in e-mail capability to communicate availability of course materials to students. By controlling the electronic release of materials, there is structure in the teaching/learning process.

I have used Blackboard for years in my teaching, because it is one platform often provided by institutions of higher learning to supplement brick-and-mortar teaching and to connect students to digital instructional content. Typically, when integrating Blackboard in course instruction, I use it as a repository for all files associated with a given course, such as reading materials, assignments, media files, and so forth. Therefore, Blackboard serves as a portal through which students access content relevant to course modules. It is a relatively easy system for students to use, because it is browser-based and can be accessed on most computers with the appropriate web browser.

When using Blackboard, I make *only* materials that are relevant to course instruction available via the e-learning system, and I make them available at particular times during the term, when access is necessary for learning. By implementing this kind of control, the e-learning system is integrated into the whole learning experience, and students can rely on the system as a consistent instructional component. Information about and for the course is posted and available when I plan it to be, and, in terms of pedagogy, when it should be, not necessarily when students want it to be. With the tool Blackboard, I create organization in the educational trajectory that is oftentimes missing in remote instruction, and I lessen the likelihood of a disorganized learning experience. Without the tool, the *willy-nilly factor of mobile education* prevails.

> For example, during one term with a particular course, I conducted an experiment. I did not control student access to the files. I did not set the date and time availability. Instead, I posted all of the materials during the first week of the course, at one time. I had organized all individual class files in folders that were ordered chronologically and clearly marked as to relevance to class instruction:
>
> Week No. 1/File Folder No. 1,
> Week No. 2/File Folder No. 2,
> Week No. 3/File Folder No. 3, and so forth.
>
> However, students could access, view, and read the files on their own schedule. And they did. Most students accessed the information in a random manner, downloading and opening some or all of the files when they wanted to, not when they were supposed to. This created a trajectory that was absent a beginning, middle, and end. Even though I stressed to the students, in e-mails and in message postings on the Blackboard home page for the course, that any information not relevant to the course modules should be disregarded, and that no information should be accessed until formal class instruction required access, curiosity got the better of the students. They opened files willy-nilly. Access to information was on their terms, not on mine, and a disorganized, unstructured learning experience prevailed.

Because students in the previous example had opened and viewed files for future modules when they were not supposed to, students knew what was coming. Many had difficulty focusing on what was being taught, because they anticipated what would be taught. This was not true across the board. Some students abided by the instructions. However, for the most part, I found that, if the files were available, they would be accessed and opened. Therefore, the experience of education had evolved into a curiosity-seeking mission, instead of a methodical process. The educational trajectory was, as stated earlier, directionless. And the intended learning outcomes were not realized.

I faced a dilemma. In the experiment described earlier, curiosity had compromised method and made for chaos in the students' education. Yet, curiosity is so important to learning. *Seek and ye shall find. Turn over a stone and you may find a treasure*. Without curiosity, humankind would not have sailed across oceans, advanced cancer research, written epic novels, or invented tablet computing. We must keep curiosity in the mix. But, as curiosity applies to course instruction with mobile technologies and distance learning, we must guide it with an awareness of context.

Education with mobile technologies and online platforms is, for the most part, *education on the move*. It is education that is convenient to populations of students who have full-time jobs, are parents, live great distances from campus, or who cannot otherwise commit to site- and time-specific class meetings. This kind of education serves a good purpose, by making education available to demographics of students, who, without mobile education, may not have any education. But in order for convenient education to produce quality learning—so that curiosity does not give way to chaos—instructors must organize and structure the mobile and online presentation of educational content methodically and carefully.

In a brick-and-mortar classroom, where course instruction occurs in a set place and time period, the regularity of place and time provide structure. With mobile and remote education, it is in the interest of instructors, as well as students, to determine precisely when students will engage with educational content and learning information in order to provide structure. As already mentioned, some of this structure can be created by date and time control mechanisms built into e-learning systems. Controlling access to electronic teaching materials can also be done with specialized client-side software, as well as with server-side controls on the computers storing the instructional data. With remote instruction, both pedagogical and technological frameworks must be solid. Interestingly, when relying on date and time controls, it is the technological framework that makes the pedagogical framework solid.

With mobile learning, the student is a user, and the effectiveness of the user experience, as created by new media technologies, is important in determining how the student engages with the learning process. The more a student is engaged, the more the student absorbs and learns. Devising and implementing facile user experiences are paramount when educating students with mobile devices, remote systems, and online platforms.

> For example, in an online course where students are scattered about in various geographical locations, the instructor's presence is likely represented by a series of hyperlinks on a webpage. There is little, if any, one-

on-one guidance, focused tutelage, or spontaneous exchange of ideas. The meaningful responses to individual questions take the form of e-mails or online postings. And more importantly, the human connection is lost. There is no glance with supportive eyes that communicates *you got it*, when a student gets it. There is no positive reflection in a voice saying *that is correct*, when an answer is correct. There is just a computer screen emitting light.

Contrast the user experience that is human and emotional in a brick-and-mortar classroom with that which is binary and antiseptic in a digital environment. When you do, you realize that if we take the humanness out of education, we run the risk of creating a digital inferiority complex that will affect students in different ways than making them feel self-conscious for not owning certain technology. In this instance, a digital inferiority complex is not due to a lack of funds needed to purchase equipment, but instead, the complex is rooted in the individual student's inability to properly connect to the experience of education.

Whenever we can streamline the teaching/learning process with mobile education, we should. This includes selecting technologies and devices that are easy to use—again, an emphasis on intuitive. By eliminating a steep learning curve with the technology, students are more likely to focus on the course material and less on how to work with the device. That is not to say that devices used for mobile education, like smartphones and tablet computers, do not have learning curves. They do. And the learning curves vary from manufacturer to manufacturer, as does the logic for navigating around different operating systems and accessing data in various graphical user interfaces. However, once the logic of a technology is understood and the fundamentals for using a device are learned—neither of which takes the college-age demographic very much time or energy to do—mobile education begins, and the learning process can be productive.

With advances in new media technologies, mobile education reaches beyond browser-based e-learning systems and online classes. Presently, educators are incorporating the use of mobile applications for handheld devices in their lesson plans. These applications, devised mainly for smartphones and tablet computers, present content in an interactive environment for users of all ages—toddlers to senior citizens. For the most part, mobile applications are intuitive. Some are so intuitive they seem magical. Swipe a finger across a touch screen and access content, flip pages, make digital art, edit video, and create spreadsheets. When a tool is that easy to use—no awkward pushing of keys, no scrolling through long webpages, and no repeated clicking of a mouse—it is more than intuitive. It is emotional. This is good for the educational experience.

When choosing an application to support curriculum, the success is in the details. Specificity is important in the development and deployment of educational applications for handheld devices, and the number of applications available for download is staggering. There are applications for everything from games that teach academic subjects, to interactive learning experiences based on GPS coordinates, to multi-user forums where students can participate in the learning process in real time with their peers. And in each of these categories, there are niche and hyper-niche topics to discover.

We need to look long and hard at not only what the application aims to teach but how the material is being presented. For example, if you teach history, you will need to decide if the mobile application should have high-definition maps, scanned historical documents, archival videos, educational games, or searchable quotations from persons of historical significance. Some of these items may contribute meaningfully to learning outcomes, and others may not. We need to do ample research into which applications are right for teaching, by actually using the applications before integrating them into pedagogy. This takes time and costs money, and it needs to become part of the course development process. The challenge here, like with so many uses of technology, is that obsolescence is always just around the corner. New applications will be available before we have the chance to fully integrate the ones we are working with into a course or curriculum.

Applications for mobile devices are designed for relatively short visits—meaning users want information quickly and easily and will not stay with a device for a very long time. This is partly due to a shortened attention span in the age of new media. Generally, users of handheld devices are on the move, multitasking, or operating more than one device at a time—a smartphone in one hand, a mobile game system in the other, an MP3 player's earbuds inserted in ears, and a tablet computer nearby. This brings up the basic question, not yet raised: Does the intermittent and transient use of handheld devices support using them in education *at all*? It also brings up the following question: Knowing that multitasking is not optimal in education, why would we require a device in mobile education, like a smartphone, that students will use for sundry tasks simultaneously, and likely, along with other devices? The answer to both of these questions comes from this book's second chapter, "Technology, Purpose, and Meaning":

> Guided by instructors who balance theory [*why*] and practice [*how*] in pedagogy, who support the innovative application of new media, and who encourage students to think out of the box, students are more apt to use technology in new ways and produce meaningful digital works that work.

It is not to say that *why* is more important than *how*. Balancing the two is the real goal. In short, make sure the pedagogy that is developed for mobile education works *with* the chosen devices and *for* the students using the devices and that students clearly understand the role the devices play in their education.

When mobile learning does not provide a facile user experience, the student is not immersed in the educational process as much as he or she is tethered to it via online forms, hyperlinks, animations, and video tutorials. The mobile experience, with its electronic connections, is, by its technological nature, less immersive than the brick-and-mortar experience, with its human connections. If the user experience is clumsy, awkward, or unintuitive, mobile education becomes even less immersive, resulting in a substantial disconnect between the student and the course material. With mobile learning, when there is little or no emotional connection to the experience, the student feels and believes that the mobile education has less value. This results in a student who may not be very motivated to learn and, therefore, does not learn. Because of this, a student is left to feel insecure or inferior.

Some proponents of mobile education may find creative ways to simulate immersion and boost motivation with virtual communities and other digital environments, and there are lots of tools to help in this effort. Skype is a popular and free software tool that allows users with a web camera and a microphone to communicate via their computer to other people by voice, video, and instant messaging. Adobe Connect is a web conferencing platform that provides users the ability to share a variety of content types, including multimedia files, in addition to networked video conferencing via the Internet.

However, virtual immersion will never be as deep or authentic as emotional immersion. With mobile education, the student does not have the kind of visceral connection that will involve him or her with the material to the point of making the acquisition of knowledge long lasting, yet. All educators would agree that long-lasting knowledge is an important and necessary end to successful teaching. Therefore, in learning, we cannot discount the importance of emotion, which is both beneficial and relevant in our lives. In an article entitled “The Other Education” *New York Times* columnist David Brooks writes the following: “Our emotional educations are much more important to our long-term happiness and the quality of our lives.”[23]

Emotion connects all of us to experiences in our everyday lives and makes those experiences rich, memorable, and meaningful. Our human condition values experiences that are emotional, because they are impactful. Emotion does the same in education by signaling there is something at stake in the process, that learning matters, truly matters. In his book *Think Better: An Innovator's Guide to Productive Thinking*, author and inspirational speaker Tim Hurson states the following: “We all have the potential to think better, more creatively. What we need is the incentive.”[24] Educators can support an emotional connection to education by creating incentive in students to learn. Creating incentive with technology is possible, when technology is used thoughtfully, purposefully, and systematically. Educators create incentive, when we persuade and convince students to use technology in order to engage with content, and we guide students in their engagement—with both structured learning plans and good old-fashioned encouragement.

In this book, I have stressed the importance of student engagement and how technology can facilitate it. However, technology can also compromise education when students rely on those gadgets, gizmos, and platforms that reduce emotional connection to the learning process. Certain technologies will not support a meaningful and emotional experience, similar to why they do not support a facile user experience. The technologies may be clumsy and unintuitive. Also, technology can get in the way and be an impediment to engagement, and thus to learning. Mobile education—with its slick front-ends, complex back-ends, and sophisticated handheld technologies—can distance students from the educational process by placing them on the periphery of learning, as they work to figure out how certain technologies work and what to do with other technologies. When students are on the periphery of learning—observing it and figuring it out, more than connecting with it and

[23] David Brooks, “The Other Education,” published November 26, 2009, http://www.nytimes.com/2009/11/27/opinion/27brooks.html

[24] Tim Hurson, *Think Better: An Innovator's Guide to Productive Thinking* (McGraw Hill, New York, 2008), 7.

participating in it—they feel alienated from the process. Education has less value.

Another important issue for institutions that are considering distance and/or mobile learning is the perception that distance learning is an inadequate form of education. In a recent poll conducted by the Pew Research Center, "just three-in-ten American adults (29%) say a course taken online provides an equal educational value to one taken in a classroom."[25] Considering that the course material in the online version of the course used in the Pew study was identical to that used in the classroom version, this statistic is telling.

Mobile education is good for an institution's fiscal bottom line and a good way to raise needed revenue, because the overall cost to produce and distribute mobile instruction is lower than with brick-and-mortar education. As a result, online education is growing exponentially. It is not uncommon for universities with distance learning programs to boast of delivering courses to thousands of students per month in hundreds of countries, with a large percentage of annual enrollment growth. The same Pew Research Center article cited earlier states the following: "College presidents predict substantial growth in online learning: 15% say most of their current undergraduate students have taken a class online, and 50% predict that 10 years from now most of their students will take classes online."[26] This, of course, is the view of college presidents polled by Pew Research, not the views of instructors, students, other administrators, or college presidents not polled—across the globe. Still, it is a persuasively high percentage, and it supports the notion that institutions of higher learning see a future in online learning. In economically difficult times, institutions of higher learning need to be both fiscally responsible and creative. Mobile education is an option they are vigorously pursuing.

There are other factors that should be put into the mix before using handheld devices to serve pedagogy—for example, the reliability of devices. Technology author Scott Weiss, in his book *Handheld Usability*, reminds us that handheld devices function mostly in "on-the-go environments where distractions are frequent, where batteries run out, and where Internet connec-

---

[25] Kim Parker, Amanda Lenhart, and Kathleen Moore, "The Digital Revolution and Higher Education," published August 28, 2011, http://www.pewsocialtrends.org/2011/08/28/the-digital-revolution-and-higher-education/2/

[26] Parker, Lenhart and Moore, "The Digital Revolution and Higher Education."

tions are 'spotty.'"[27] Therefore, relying on handheld devices that have unreliable parts, faulty energy sources (weak batteries), and intermittent Internet connectivity, could prove disastrous in teaching. Students' focus will inevitably come and go, as batteries die and Internet access pops in and out.

The issue of time is especially important. Things happen a lot faster in technology than in education. In technology, months can seem like days, days like minutes. This will necessitate making swift decisions about tools and platforms and trying them out quickly. However, in higher education, where there can be committees of people who need to meet and vote before certain technologies and related components can be incorporated into curricula, time passes slowly. I speak from experience. Several times in my academic career, I have been a member of university faculties tasked with devising new curricula that require new media tools, technologies, and platforms. This is a long, sometimes complicated process, where there can be many levels of approval required before even a single course or tool is adopted.

So, hardware, software, Internet connection, topic specificity, and time constraints are items to consider when integrating handheld devices and mobile applications into pedagogy. With all of this, there are also budgetary considerations for programs and institutions, as well as for teachers and students. An important question needs to be answered: With all the practical and pedagogical considerations that need to be addressed, is there enough advantage for teaching with handheld devices and mobile applications that the monetary investment in them is justified? The investment of money cannot be underestimated. With these technologies, like with so much of new media, there is a real need to upgrade often—the constant inconstancy of new media. This can be expensive.

There will be communities and sectors of society where students and teachers are not able to afford the expensive technology—either the initial purchases or the constant upgrades. This is a recurring issue with new media in education, illustrated by the *app gap* between rich and poor students, discussed earlier in this book. However, this issue has special resonance when discussing handheld devices, because there is an assumption that they are standard across demographics and that smartphones are communication fixtures owned by everyone. This is not true in all of the United States and across the globe. Technological elitism, also discussed earlier, is substantial

[27] Scott Weiss, *Handheld Usability* (John Wiley & Sons, Ltd., New York, 2002), 20.

in the world of handheld devices and mobile applications. We cannot assume that everyone has a smartphone.

Mobility in education is filled with uncertainty. The uncertainty comes from the fact that professional educators have not used these devices and applications in teaching for very long. There is little data that show the long-term benefits of using these tools in the classroom. So, we must wonder *if* there is any long-term benefit. I seriously question if the mobile education model we know today is sustainable long term, not in terms of economics, but in terms of the level of emotion needed for successful learning. (I will, however, add that it is a model not fully formed or fully tested at this time. Because of the ever-changing state of new media, how could it be?) When one-on-one guidance, spontaneity, and *you got it* are missing from the educational process, we run the risk of creating education that is standardized and utilitarian. With education that is standardized and utilitarian, students focus on memorizing data and learning rote formulae, as opposed to experiencing education.

Moving forward, a process of trial and error will determine if handheld devices, and the cornucopia of applications we can load onto them, will redefine mobile education in any significant way. Let us not impose standardization and utility on mobile education. Instead, let us be inventive with new technologies and make rich, emotional learning experiences for our students. The magical intuitiveness of these devices and their ability to captivate us with wide-ranging content may be reason enough to use them.

13

INTERDISCIPLINARY IDEA EXCHANGE

New and interesting combinations of interdisciplinary study are necessary in higher education in the 21st century, because the professional worlds we are preparing students for are varied and interconnected. New media technology is fast becoming a bridge that connects disparate disciplines: healthcare with arts, media with natural sciences, journalism with design, and the list goes on. In order to meet the social, economic, educational, and cultural needs of society in the future, institutions of higher learning are creating interdisciplinary areas of study that will prepare students for specialized professions. The educated person who will achieve success in this mash-up of disciplines is the creative thinker, the visionary who can think out of the box and produce innovative work in collaboration with other people in various disciplines.

Much of the global economy is shifting away from manufacturing, as we have historically known it, and toward arenas where creative output is the focus—arenas such as design, culture, environment, and science—in which success seems to depend on innovation. In the classroom, it is difficult to teach students innovation per se, but we can prepare students for interdisciplinary arenas of employment by exposing them to new concepts, encouraging them to think creatively, and teaching them to make unique works. New media is ideal for this. Students, who work with technologies and tools that change all the time, will be more adept as professionals at thinking spontaneously and coming up with creative solutions to problems than those who do not. Knowing the *how* and the *why* of using new media technologies—that is, pushing buttons with purpose—creates unique experiences in the classroom and prepares students for the unexpected and unpredictable situations that may await them in the interdisciplinary world.

The students we educate to be the leaders of tomorrow must be able to organize ideas, brainstorm concepts, plan projects, produce content, negotiate deals, and execute plans in diverse, multifaceted, and oftentimes unpredictable settings. Our students need to be innovators, not just locally but globally. Success in and with these interdisciplinary arenas, which I call *inter-disciplines*, depends on it. Because new media is everywhere, they influence how messages are communicated, creative visions are expressed, and products (both tactile and cerebral) are made in cultures around the world. The student who can master the effective use of new media tools and technologies in order to reach diverse peoples and affect change in many nations—whether that change is political, cultural, or economic—is a new, creative thinker. This is a thinker who excels within an interdisciplinary framework built by new media, has tools proficiency, and thinks anew when utilizing that proficiency. This is *new creativity*.

Seth Godin, author of *Linchpin: Are You Indispensable*, emphasizes the importance of new creativity in our everyday lives, and he provides an explanation for this new thinker. Godin encourages us all to think beyond the limits we set for ourselves and others set for us. He expects us to rely on imagination to devise solutions to problems large and small, and in our daily lives, he wants us to innovate, innovate, innovate—to the point of being artists. Godin believes there is an artist in all of us. We just have to nurture that artist. To Godin, the artist is not necessary a painter, filmmaker, singer, dancer, cellist, or pianist. It is a new kind of thinker.

> I think art is the ability to change people with your work, to see things as they are and then create stories, images, and interactions that change the marketplace.[28]

> Your art is what you do when no one can tell you exactly how to do it. Your art is the act of taking personal responsibility, challenging the status quo, and changing people. [29]

---

[28] Seth Godin, *Linchpin: Are You Indispensable?* (Portfolio/Penguin Group, New York, 2010), 91.
[29] Godin, *Linchpin: Are You Indispensable?*, 97.

Godin's thinker is someone who can work within various disciplines, affect different peoples, drive diverse markets, and understand distinct societies. This thinker will be the most creative artist, the most productive student, and the most successful professional in the new, creative economy. I believe this kind of thinker/visionary is the prototype professional that educators should prepare all students to be. And we can do it with new media.

Because new media provides students, and all of us, the opportunity to innovate globally, new media is important if we are to realize Godin's mandate to "challenge the status quo and change people." The use of social media networks to organize people and communicate messages during times of crisis, such as during the 2011 uprising in Egypt known as the Egyptian Revolution, is an example of new media's power and reach. As the crisis unfolded, Western countries became aware of the Egyptian uprising on social networks like Twitter, and they were motivated to assist the Egyptians, who demanded the ouster of their dictatorial leader. Another example is the protest movement Occupy Wall Street, which relied on the web, social media platforms, and new media devices, to mobilize activists from New York to Chicago to San Francisco to cities around the world—with a mission of bringing awareness to economic inequities and the perceived misuse of power by corporations. The sheer numbers of people—of different ages and from different socio-economic, ethnic, and professional backgrounds—brought together by technology for both of these causes, are exemplary of new media as an interdisciplinary, international, cross-cultural force that enables us to challenge the status quo and change people.

The interdisciplinary potential of new media is recognized in higher education as an important trend, so much so that new media is integrated into university curricula in disciplines where it would be expected, such as electronic media and interaction design, *and* in disciplines where it would not be expected, such as natural science and literature. In addition to the academic value of this trend, there is practical benefit. Using new media tools across academic disciplines helps justify the expense and defray some of the cost of the tools. It is not unusual for certain institutions to have one computer lab, outfitted with state-of-the art hardware and software, serve several departments, schools, or colleges. More common is a central lab, in a library or other general use building, where students across campus can work. There may also be facilities where students can borrow equipment such as cameras, laptops, or recorders, checking out what they may need for a class or an interdisciplinary project. This model of a single lab serves a broad educational mission—it brings students from various disciplines together—as well as an institution's bottom line—it saves money. There is shared vision and shared cost.

Because new media is the bridge bringing these disciplines together, in the lab and elsewhere, new media influences the evolution of these individual disciplines.

> For example, if in a geology course, a student uses Adobe Flash to create an interactive timeline that illustrates how fracking (the process of fracturing a rock layer with a pressurized liquid to detect the presence of natural gas) unfolds from beginning to end, the student has created a way with Flash to document a process in geology. Since interactive timelines can be effective in documenting processes in other disciplines—such as medicine, art, chemistry, or literature—students in other disciplines should take this Flash course. If the same Flash course were to also have chemistry, art, and literature students in it, all of them would share ideas, learn about each other's fields of study, and collaborate on interdisciplinary projects, in the same classroom environment and with the same tool. The tool becomes a means to interdisciplinary learning.

Given that we are a connected society—in terms of the Internet, e-learning systems, and social media platforms—we share ideas easily and freely. Whether in or out of the classroom, we are exposed to new ideas, academic trends, and creative waves simply by turning on a device. Because of technology, there is a profound *idea exchange* of many different topics and issues, very quickly. Sometimes ideas are exchanged while learning a new technology, as in the previous example. Sometimes ideas are exchanged when networking with technology, as in the case of social media. Sometimes ideas are exchanged by mobilizing people to get involved with civic or political causes. The bottom line is that there is a mix of ideas across disciplines, and the spread of these ideas is far-reaching and vast.

With all the new creativity buzzing around our electronic communications systems, how do we synthesize these many ideas into coherent communication, so that collaborative vision is focused and the idea exchange is innovative, not convoluted? How do we filter out of an inordinate amount of information what is truly necessary to the exchange and what is not? We simplify. We break down the idea exchange into four basic stages:

Four Stages of the Idea Exchange

Stage No. 1: Communicate

The information in the exchange must be communicated in an organized, cohesive, and intelligible manner, from sender to receiver.

Stage No. 2: Receive

There needs to be confirmed receipt of the information by the receiver, in order for the exchange to be valid.

Stage No. 3: Process

In order for the idea exchange to be consequential, the receiver needs to evaluate the relevance of the information.

Stage No. 4: Inspire

When the information is deemed relevant, the receiver will be inspired by it and will create new ideas.

Inspiration is key to the idea exchange. Inspiration is what produces original thought. In an interdisciplinary context, this means coming up with new ways to think about, study, and work in hybrid arenas, as well as arenas that have yet to be defined. New ideas result from inspiration. Inspiration results from a process of clearly communicating messages. Going back to Seth Godin's definition of a creative thinker or artist, we learned that "art is the ability to change people with your work."[30] Clear communication of message is necessary if you want to "change people." Quite simply, your message has to reach people, or your art will have no effect on them. If people do not receive your message, they will not be inspired. They cannot be changed by something they never got.

On the Internet and in the vast world of new media, it is not easy to communicate message clearly. There is a plethora of data out there and a deluge of communication coming at us in many formats: as bits of data

---

[30] Godin, *Linchpin: Are You Indispensable?*, 91.

strewn across blogs, in wikis, on a vast array of websites, as text messages, tweets, postings on electronic message boards, *likes* on social media networks, ratings on online retail platforms, and so forth. Some of this electronic information coming at us is useful. So much of it is not. Yet, we must work through it, read it, listen to it, interact with it, and assess it, in order to ascertain if it has any value for us. And we have to do this constantly. Much of the information is sent with a deliberate agenda: persuade, sell, entertain, teach, and so forth. The information does not always inspire its recipient. It cannot *always* inspire. This saturation of electronic data that has little meaning or value to us, and is around us always, is *digital noise*.

In his book *White Noise*, author Don DeLillo uses the theme of media saturation to underscore a series of devastating occurrences to a family at the center of the book's plot. The media—the *noise*—are pervasive in the characters' lives, to ill effect. The family lives "in a high-technology society: there is abundant information around, but nobody seems to know any thing."[31] This is similar with digital noise: There is abundant information around, but nobody seems to know *any thing*. The digital noise seems omnipresent and palpable, because we feel it. When digital noise disrupts communication, the idea exchange fails to inspire others. Digital noise is invisible, but we are aware that it is invading our *personal data zone*.

A personal data zone in new media is similar to personal space in psychology. This is an area, a space, that we regard as our *own*, and when this space is encroached upon by any number of factors, we feel uncomfortable, even troubled. These factors include other people's smartphones ringing at inopportune times, such as during an exam, movie, or funeral; a news app bleeping on a mobile device every time there is breaking news; and even music blaring on someone else's earbuds, especially if you are sitting next to that person in a car, on a train, or on a bus. Psychologist Edward T. Hall, author of the book *The Hidden Dimension*, qualifies the reaction of people to the invasion of their personal space or "small protective sphere" as one in which a person feels discomfort, anger, or anxiety. [32]

The reaction to the invasion of our personal data zone is similar to that of the invasion of our personal space. All of the communication, messaging, and stimuli we are exposed to on a daily, hourly, and minute-by-minute basis

---

[31] Mark Osteen, *Don DeLillo—White Noise: Text and Criticism* (Penguin Books, New York, 1998), 3.

[32] Edward T. Hall, *The Hidden Dimension* (Anchor Books, New York, 1966), 112.

can create undue stress, because, if we do not want to receive it, we are forced to deal with it. Or, in the case of wanting or needing information from the web, we are forced to filter it. We have to, in order to make sense of it. In order for the data to be useful, we must extract the noise from it.

What mechanism or system allows the thinker/artist in all of us to filter what we need for inspiration and discard what is extraneous? Ironically, the answer is *new media*. New media technologies can be used effectively to filter out inconsequential information. With new media, we can choose the information we want to allow into our personal data zone. In one very basic example, an e-mail filter can keep unwanted spam out of an inbox. Using new media technologies to filter what information we keep and what information we delete will help make the idea exchange consequential. This will lead to inspired interdisciplinary thinkers.

For educators, filtering out digital noise is simply a matter of setting clear objectives for the use of tools and technologies. In order for students to learn, to benefit from digital data, and ignore digital noise, we need to instruct students in how to receive information that is *useful* to producing new creativity and to disregard information that is not useful. Create exercises that teach students to prioritize data types, so *they* can determine what information they should let into their personal data zones. Priority data is that which begins as organized, cohesive, and intelligible communication (stage 1 of the idea exchange) and ends as relevant communication and inspiration (stages 3 and 4 of the idea exchange). By eliminating digital noise, the idea exchange will be more meaningful, the new ideas more creative.

It is ironic that some of the technologies we choose to keep out of our classrooms, because they are diversions, are the same ones we can encourage our students to use to filter out digital noise. Text messaging, or texting, is one example. Texting is a phenomenon that sometimes I cannot understand. I do not find texting an intuitive or facile way of communicating. Our thumbs and fingers are just too big to successfully push tiny buttons, or graphical depictions of buttons, or electronically generated buttons on a user interface. Even the most sophisticated, touch-screen smartphones, which are commercially available, require our very human fingers to push very non-human-sized letters and numbers.

The physical limitations of texting, along with the cryptic syntax of *text-speak*—which is a combination of reinvented spelling, unique punctuation, and the use of symbols called emoticons to communicate emotion in an environment in which there is no emotion—make texting a curious form of communication. Because texting is limiting and cryptic, it is a substandard way to communicate consequential messages. However, texting is a prime way to communicate digital noise. It is too challenging to communicate fully realized thoughts and too easy to communicate gibberish with texting. BOCTAAE :-)

With all that said, there is a productive use for texting in the classroom. It just takes some serious thinking beyond the screen. Indeed, texting can be used effectively in teaching, as long as its use is repurposed *for* teaching and instructors work *with* its technological limitations. Similar to the repurposing of Twitter in "Case Study No. 3: Engaging Students with Twitter," rethinking how texting is used and giving it a new purpose can be useful in education, specifically in research, as I discovered in the following case study.

Case Study No. 5: Conducting Research with Text Messaging

This case study focused on the use of text messaging, or texting, to conduct research for a formal research paper. The objectives of the case study required students to study a particular international news topic: the U.S. war in Afghanistan from 2001 to 2012. During classroom instruction, over a two-week period of time, students were engaged with lectures and discussions that did the following:

1) Presented the wars as unresolved conflict of international significance.
2) Explained the history that preceded the conflicts.
3) Documented the current status of the war.

The main objective of the study was to use texting on mobile devices to accomplish the following:

1) Aggregate information in the form of a *text string*—a series of statements and responses—that would constitute an outline for research.
2) Use this outline in a brick-and-mortar library, as a guide for obtaining additional information from filtered resources that would support a thesis for a research paper.

3) Write a formal research paper that proposed a way to resolve the conflict.

Because privacy is a very important issue to me, when using digital devices in and out of education, I asked students to volunteer for this study, as I would be acquiring their phone numbers when they texted me during the study. Only students who agreed to participate did. Interestingly, over 90% of the students opted to participate. Expressly for the purposes of the case study, I would create a single list of contacts using the phone numbers I received. This would constitute a small database of contacts called *contacts database*. This one database, which contained a comprehensive list of the students involved in the study, would help me streamline communication to student participants via text messaging. Students who agreed to participate in this study knew they would be included in the contacts database.

This study began with each student texting a single message to me, in the form of a question about the U.S. war in Afghanistan. From the more than 60 questions sent to me, I selected the following two:

1) How are the children of Afghanistan being protected during the ongoing conflicts?
2) Will the death of Osama bin Laden help to end the war sooner than expected? (It should be noted that at this point in time, the U.S. government had already reported that American soldiers had killed Osama bin Laden.)

I then asked students in class to choose one of these two questions, as a research topic for the entire class, by voting with a show of hands. Of the 54 students participating in the study, 36 opted to research topic number 1, "Children and Aid Organizations," while 18 opted to research topic number 2, "The Future of the Afghanistan War without Osama bin Laden." Number 1 would be the research topic for this exercise.

Conducting the study with two questions simultaneously would not work, because a single, uninterrupted text stream was required. I proceeded with the case study, and the research process via text messaging began.

Research Objective (Text Message)

I sent all students in the contacts database the following text message to begin the research project:

- To address the matter of how the children of Afghanistan are being protected, research NGOs involved in the humanitarian effort, choose one, and provide the following information:

    a. Name of NGO.
    b. Mission of NGO in one sentence.
    c. Two positive results that have come from the efforts of the NGOs.

Research Data String (Text Message Replies)

Students were instructed to provide the requested information (above) in a text message by *replying to all*, so that all members of the contact database would receive the same research data.

Because text messaging was at the core of this case study, it was important to use standard texting techniques—that is, short sentences, pithy phrases, abbreviations, and so forth. This would add authenticity to the study and enable me to better determine whether texting is a worthwhile means of conducting research or at least a tool to assist in that effort. However, I could see from the beginning, from doing a preliminary test exercise, that both brevity of message and the cryptic language style of text messaging might be issues—in that texted communication can sometimes be incomplete and/or open to interpretation. To compare texting language with standardized English language, and to illustrate the challenges of communicating with texting in research, I have provided a text message sent by one of my students in this study, along with a fully spelled out version of the same passage, following.

Student Text Message for Research Project (Received)

The text below is an actual text messages received from one of the students.

- i c many issues with the humanitarian effort rite now bc some rebel groups won't allow ngos access to villages. iaw the afgnstn govts policy, ngos from approved orgs such as un, who and red+ should have access to provide aid. b4 kids can get help, there needs to be stricter enforcement of policy so ne1 who wants to help the ppl of afgnstn, young or old, can.

Student Text Message for Research Project (Spelled Out)

The italicized text below shows which words and terms in the previous message were abbreviated using text-speak.

- *I see* many issues with the humanitarian effort *right* now, *because* some rebel groups won't allow *NGOs* access to villages. *In accordance with* the *Afghanistan government's* policy, *NGOs* from approved *organizations* such as *the United Nations*, *the World Health Organization* and *the Red Cross* should have access to provide aid. *Before* kids can get help, there needs to be stricter enforcement of policy so *anyone* who wants to help the *people* of *Afghanistan*, young or old, can.

There is a marked difference between the message received and the one spelled out. Although the previous examples illustrate the difference between texting and standardized English, it was apparent that there would be variations in how students wrote their individual text messages. Each student would have a unique style of communicating with text-speak. For this study, and for the integrity of the research conducted in it, it was important that all students had a consistent language with which to communicate and share research data on their mobile devices. There needed to be a single text messaging language that would be interpreted by all student researchers in the same way, because the research information had to be the same for all students.

Therefore, borrowing from readily accepted jargon used at the time of this case study by the mobile technology community for texting—as well as variations thereof, such as instant messaging (IM) and short message service (SMS)—I created a single document of approved text messaging language for all students to use in this exercise. This document, which I called "Approved Research Text Messaging Language," listed the abbreviations and symbols students could use in their communication and research. Students who deviated from this approved list would be penalized in their grade. It was necessary to create a new text messaging language—one that would standardize communication—because at the time of this case study, one single language did not exist.

Working with this standardized texting language proved beneficial, generally. However, this set language was not enough to insure that all students would communicate in an absolutely standardized way. This was due to the fact that some abbreviations communicated in the exercise came from the English language, not from text messaging language. These abbreviations would also need explanation. For example, some students did not know what the acronym NGO stood for, and they did not take the initiative to find out. To avoid compromising the research effort, because some information could not be fully explained in a text message, I provided students with necessary supplemental information in class lectures. The objective here was not to undermine the texting exercise but to make certain that students had clarity and understood the research data.

Therefore, in class lectures, I verbally communicated other necessary and applicable information, such as the following explanation:

> NGO is an acronym for non-governmental organization. An NGO is typically a charity, foundation, or aid organization focused on helping the young victims of war. In our study, these victims live in Afghanistan.

Here, this case study demonstrated another hybrid approach to teaching with new media: supplementing an exercise focused on text messaging with in-class lectures. As with most of my experiences integrating new media technologies into classroom instruction for the various case studies in this book, I found here that combining technology with human engagement worked best.

Continuing with the study, out of a possible 36, I received 33 replies to my initial set of instructions: Address the matter of how the children of Afghanistan are being protected. This was a sufficient number of student responses to continue to the next phase of the case study. Following is a list of some of the 33 replies I received, in addition to the example provided previously as "Student Text Message for Research Project (Received)." The examples that follow were written in accordance with the "Approved Research Text Messaging Language," created for this research exercise.

Sample Text Replies Received from Students

Following are student responses to the initial text message from the professor, *"Research Objective (Text Message)."* These messages appear exactly as they were texted.

- WHO World Health Organization. Goal is to improve health status of Afghanistan ppl. Help girls get education, despite Taliban band. Routinely vaccinate children against disease.

- un food + agriculture org
  rehabilitate agricultural sector in afghnstn
  combat animal diseases… foot + mouth disease
  distributed tons… wheat seed + animal feed

- Red Cross Rspnds to humanitarian needs of people affected by armed conflict
  Protects detainees-
  assist wounded and disabled-
  supports hospital care-
  improves water and sanitation

The main objective at this point in the study was to create a text stream, with contributions from each of the students, which would be used as a foundation for research by all students. Aggregating the initial research contributions from the students would be easy, as students were instructed to *reply to all* to my first text message, "Research Objective (Text Message)." This way, each student would have the responses from all 33 students on their smartphones as a single stream. It was never intended that text messaging would be used as a research tool for the entire research period or during the period of time when students would write their papers, which would be four weeks.

Text messaging would provide students with a common framework with which to begin individual research. With this initial text stream in hand, individual researchers set out to do further research in brick-and-mortar facilities, such as libraries, with printed information from various periodicals and with sundry other resources. From here, the research effort for students became more individual, less communal. Supplied with a research objective from me and a research framework from one another, in the form of a stream of text messages, students were ready, smartphone in hand, to write their papers. And they did. Both their interpretation of the research outline and their development of a single thesis in each of their papers were as unique and accomplished as each of the students.

Interactive learning was part of the process. Students engaged with their smartphones in order to generate research data and sent the data via text messaging to the class stream. With the use of their mobile devices, students determined *how* the research would be carried out—with device in hand, in a library, moving from desks to computer stations, to information kiosks, and to bookshelves. This meant that the interactive learning would be mobile. As students conducted research using their smartphones, they undoubtedly learned from the material.

Participatory pedagogy was also part of the process. Because students were participants in the creation of this research methodology, which was foundational to their learning experience, they established course flow and determined how course materials would be presented, specifically for the research portion of the course. The pedagogy students participated in creating provided a framework for the research effort. It was an adaptable framework. It could work in the library, in the classroom, and even at home—wherever students had their smartphones.

Had this process been more traditional—meaning the instructor typed out an outline on paper and handed it out to students or posted a digital version of the outline (as a PDF or .doc file) on an e-learning system—it would not have been as streamlined and likely not as successful. With a paper handout, some students would certainly have lost the research guide. With an electronic file posted on an e-learning system, there would have been several, perhaps many, steps involved in students' accessing the data: logging onto the e-learning system, downloading the file, saving it to a laptop or a portable storage device (like a USB drive), and then accessing the guide on some device or a computer in the library. With the outline on their smartphones, students were *ready to research* quickly and easily and in a short amount of time.

Generally, the use of texting in research was successful, in terms of establishing parameters for research in the form of an outline. Texting streamlined the research process by making it digital and mobile, at least in the beginning. Ultimately, students typed and passed in a full-length research paper on the topic given to them. The research process that began unconventionally concluded conventionally. By expanding the interdisciplinary idea exchange with new media technologies, another hybrid approach to education, balancing new with traditional, worked.

14

THE POWER OF GAMES

In education, the discipline of game design and development—once little known in traditional classrooms—is rapidly growing in popularity. There are majors, minors, academic programs, inter-departmental initiatives, and entire colleges focused on game design and development at various levels. More and more institutions of higher learning are developing curricula focused on game design, electronic and otherwise, because games in society are pervasive, useful, curious, alluring, and available. Games engage people in many ways. Games entertain, motivate, educate, tell stories, resolve conflicts, solve problems, and even assist in healing personal ills, such as substance addiction and depression. Games exist in arenas as varied as business, entertainment, politics, science, marketing, medicine, advertising, journalism, sports, history, performing arts, ecology, and literature. Games seem to be everywhere, and they appeal to a wide demographic of players. Games know no geographical boundaries and are cross-cultural.

Another reason why games are popular in education is that the discipline of game design and development is complex, artistic, technical, and evolutionary. The discipline morphs and grows as technology changes. The constant inconstancy of new media is especially relevant here. With its constant learning curve, game design and development is a contemporary discipline that keeps education interesting, topical, and moving forward techno-creatively, while requiring both instructors and students to stay current with the latest design and production concepts and techniques. Balancing *why* and *how* is important in this discipline, because, in order for a game of any genre to work, both front-end and back-end components must integrate seamlessly into a single game experience—whether single player or multiplayer, tactile or virtual, short or long in duration, with or without narrative, or simple or leveled in terms of gameplay. The visual aspects of an electronic game must be as appealing as its programming is functional as its experience is unique.

And yet, another reason why games are powerful in education is the same reason games have taken business by storm: profit. In business, game design and development has been one of the most profitable and fastest growing industries in history. The popularity and profitability of games in business drives the popularity and profitability of games in education. There is a lot of money to be made educating students in disciplines related to game design, production, and management, because there are good, well-paying jobs awaiting students, who have solid technology skills, design prowess, storytelling talent, and new media acumen, and who can work in a collaborative environment.

Like so many disciplines focused on technology, game design and development is susceptible to industry trends: new iterations of equipment and hardware, the use of one programming language instead of another, the ongoing development of software, and industry domination of particular proprietary gaming systems. As I have addressed numerous times in this book, the cost of staying current with industry trends in technology is high generally. The cost in the games world is higher, because consumer trends—in terms of what games people buy and the frequency with which they buy them—determine to a large extent what kinds of games get produced. Trends in the games industry are swift to come and go. Console games like Microsoft's Xbox, motion-controlled systems like Nintendo's Wii, and online game simulations like Zynga's FarmVille, have all been very popular. They may continue to be popular, or they may disappear altogether, as yet another technology is developed to make the game-playing experience more intuitive, interactive, immersive, and rich. Industry trends determine what kinds of courses are developed and offered in game design and development programs in education, and they should.

In terms of creating a curriculum that works, game design and development is challenging. Pedagogy that supports game design and development needs to be adaptable. Methodologies and materials need to be updated and replaced as needed, similar to what I discussed at the beginning of this book in the chapter "Web 2.0 and New Education":

> With new media, a successful pedagogy is an evolutionary one ... as pedagogy changes with the advent of new technologies over time, it is understandable that the tools we teach with will come and go. Therefore, it seems that our pedagogic concepts and teaching methodologies will come and go as well.

Pedagogy that supports game design and development has to be properly funded. In business, there is a bounty of investment dollars available for developing the next great game concept/system/experience. Also in business, the funding necessary to keep design and development processes active and current comes from various sources: game companies themselves, venture capitalists, and, in the case of a publicly traded company like Zynga, shareholders. In education, funding for game design and development programs, like other expensive ventures, is harder to come by.

Again, to stay current with technology and design practices, periodic infusions of cash may be required. Unless an educational program is funded by private enterprise, such as a prominent game company, or has a sizeable endowment, or receives some other kind of internal funding—for example, from a private donor or some discretionary fund—budget objectives will be difficult to meet. Without proper funding, educational programs focused on game design and development will likely cease being competitive and fall out of favor with faculty, who understand the need to be current, and with future students, who want their education to be relevant to their chosen profession in games.

In addition, the competition among reputable institutions looking to attract good students is steep, as there are top universities around the world all wanting to recruit from a limited pool of talented and capable students. Individuals in this talent pool are more likely to attend a properly funded, state-of-the-art program, more than they are a program with obsolete technologies and old equipment. After all, the objective of any student who studies a contemporary discipline like game design and development is to hit the ground running in the professional arena upon graduation, with skills that are needed and information that is current.

The complexity and collaborative nature of game design and development are both important, when considering how the discipline is integrated into curriculum. Creating an electronic game is complicated. Creating a successful electronic game is even more complicated. Creating a successful curriculum, which instructs in the creation of successful electronic games and stresses the theory/practice balance necessary in the many aspects of the industry, is most complicated. To simplify the task of creating a successful curriculum in game design and development, let us divide the field into two categories: *specific* and *general*. And, let these two categories each constitute a pedagogical approach.

With the *specific* approach, there is a single focus to a program, such as design, programming, production, *or* management. This kind of program prepares students to do one thing well: to design, program, produce, *or* manage. The curriculum is robust. Students need to master their chosen area. Therefore, there must be a sufficient and proper number of courses offered in the curriculum so that students can achieve the necessary level of proficiency. This may require creating individual tracks or sequences of courses. It would almost be like having four separate major disciplines: game design, game programming, game production, and game management. Students would choose one.

With the *general* approach, pedagogy supports a generalist approach to educating students. This kind of program provides an overview of design, programming, production, *and* management, and it prepares students to be professional generalists. Students in a *general* program are more chief cooks and bottle washers of game design and development than experts in any one area. They are able to perform in more than one position or a combination of positions—designer/manager, programmer/producer, producer/manager, and so forth—but not usually with sufficient proficiency to be a master in any one. Programs that opt for a general approach would likely prepare students to be stronger producers or managers, instead of designers or programmers. The latter positions require a higher and more specific level of expertise and more focused study. Producers and managers can be successful with a general understanding of a number of areas.

Devising game design and development curricula can be challenging. In addition to design and programming, there are many other items to consider. Here are some of them:

1) Delivery systems: mobile, console, online, and so forth.
2) Development motivations: design-driven, technology-driven, market-driven, and so forth.
3) Dimensions of game worlds: physical, emotional, environmental, ethical, and so forth.
4) Multimedia production: video, audio, graphics, and so forth.

5) Genres: action, strategy, role-playing, puzzle, and so forth.
6) Game types: educational, serious, political, entertainment, and so forth.
7) Narrative: linear and non-linear.
8) Game engines: open-source and proprietary.
9) Game balancing: a level of difficulty that is appropriate for a game's target audience.

Add to the previous list the need to educate students about demographic research, licensing, and other business concerns, all necessary for a well-rounded, general education in game design and development.

A worthwhile program in game design and development should include courses that prepare students to enter the professional games world as viable, versatile, and skillful professionals. Success in the professional games world may not be as straightforward as choosing one approach, specific or general, over the other. Success may require educators to customize programs that—in order to meet industry demands and stay current with professional trends—are at times specific, at times general, and always flexible.

> For example, if the games industry is primarily focused on online games, the curriculum should include instruction in programming languages, production methodologies, and design practices that are supported by and work in commercial web browsers. If the industry is focused on high-end console games, the curriculum should include instruction in design and coding for game engines, the software systems used for the creation and development of console electronic games. If a program determines that its students will perform best in managerial positions, the curriculum should include courses in business or even a business minor. Certain programs may find that a combination of the previous approaches works best.

Few undergraduate degree programs can do everything necessary to prepare all students for a career in game design and development, because the discipline is so vast and varied. Universities have to cover a lot in four years, the typical length of an undergraduate program. Typically, standard undergraduate degree programs include a required sequence of courses in a major discipline (the major), sometimes a minor discipline, as well as a core of required courses in various academic subjects. Usually there are too few slots available in an undergraduate student's schedule to offer the number of courses and depth of study necessary to provide a high level of proficiency in any one aspect of game design and development, especially design and programming. Some programs get around this by extending the program one additional year. This can help. But all things considered, a general approach is most practical for a program—undergraduate degree program or certificate program—that plans to prepare students for continued study, beyond undergraduate school.

The opposite is true at the graduate level, where mastering a particular discipline and/or various aspects of a discipline is expected, because of the focused nature of graduate school curricula. Generally, the objective of developers of curriculum at the graduate level should be similar to that of the games industry itself: adapt quickly and consistently. Develop a tiered curriculum that spans years, with ample time and coursework, so that students can develop expertise in more than one aspect of game design and development. For example, structure a three-year program with one year of design, one year of programming, and one year of producing and managing.

Certainly students will enter the graduate program with particular artistic and academic strengths, and during their time in the program, students will develop these strengths into fortes. Yes, students will be designers, programmers, producers, or managers first. But a successful graduate program, at the master's or doctoral level, will also provide students with a necessary secondary level of proficiency: the designer who can program, the producer who can design, and so forth. Like any complex discipline, expertise in more than one area within that discipline will give students a competitive edge.

Overall, there is good news in game design and development in education. More and more there is support at institutions around the world for the creation of game design and development programs. I can speak from firsthand experience. I have created game design and development courses at a number of universities in the United States: a well-known design school in New York City, a private institution of higher learning in the Northeast, and a large research university in the Midwest. In addition, I have taught courses related to game design in Europe, where at universities, the discipline is fairly new but catching on quickly. Currently, I sit on a university-wide committee tasked with creating an inter-disciplinary game design degree program for students across academic, creative, and technical disciplines. All of this reinforces the fact that game design and development is being taken seriously in higher education, at institutions of various sizes and in various places.

Aside from programs in game design and development that prepare students to be games professionals, we can use games to help students become better, more engaged students by using games in teaching. We can do this by integrating games in pedagogy. Why is it effective to use games in teaching? The answer is similar to why games are effective in the marketplace: user experience. Players have a vested interest in the playing of a game, because they are affected by the outcome of the game and the experience of playing the game. For example, a player involved in a serious game, whose objective is to alleviate depression, may come to understand the root cause of his depression after playing the game, or better, he or she may not be depressed any more because of the game.

Another reason is choice. During gameplay, players make choices that shape the playing experience and influence the game's outcome. Making choices empowers users. For example, in a video game, a player may be required to choose an avatar, a weapon, a location, even a wardrobe. These choices prepare the user for the journey through the game, and they will likely have an impact on how the player maneuvers through the game world and meets challenges presented in the game, ultimately impacting whether that player wins.

I incorporate games in my teaching to engage students by allowing them to make choices. These choices help determine how instruction unfolds in the classroom, and by doing so, these choices affect how students learn. By engaging students in their own learning, I connect users to experience, which is the fundamental objective of any worthwhile game. I mold the educational experience with games, so that students have the opportunity to set real, attainable goals for their learning. Students collaborate with one another on reaching these goals, and they learn from one another during this process of choice and collaboration.

Going back to several of the discussions in this book about user engagement and how effective that can be in the learning process, it is logical that teaching with games would be beneficial to students. Games are interactive. Interactive learning, as established earlier in this book, involves students in the educational process significantly more than traditional instructional methods. Because students are actively involved, they have a vested interest in their own educational process. Students are not voyeurs; they are players. Students are not attendees; they are participants. Because students are part of the process, they learn better. Also, games are immersive. Games can create mind-bending, engrossing experiences for players, for students. Using games in instruction takes interactive learning to a new level—closer to immersion. As an educator, there are few things more significant than immersing students in what they are learning. Education that is interactive and immersive is education that works.

However, the success of immersion with games in education depends on the integration of games into pedagogy purposefully, not arbitrarily. To be effective in teaching, games cannot be gimmicks for capturing students' attention, but instead, games must be vehicles for blending classroom experience, course content, and student engagement. While there are games of various genres and platforms readily available that can do this—online games, mobile games, console games, word games, and so forth—I have found that *original* games engage students more. Students have no history with new, original games. Therefore, with the creation of these games, instructors can focus on accomplishing goals specific to teaching. Original games provide for customized learning. Instructors can personalize the game experience by incorporating details of students' learning environments and addressing individual ways students learn. Because instructors can tailor learning experiences for students, original games are more effective in education than games with which students are already familiar.

With original games, students' level of curiosity about the game and their desire to participate in playing it are high. After all, the students have never played the game before. They want to know about it. They want to see how it works. They want to engage with it. The broad goal of meeting challenges in the game world becomes an instructional strategy: Engage in process and learn something. The worldview that is commonly adopted by players of a game becomes the classroom view of students: Rise to a new challenge and win. Success in gameplay is success in teaching, if the game is crafted to meet specific educational objectives. When students work to solve problems, answer questions, or meet challenges in a game that focuses on the particular subject they are studying, they learn more quickly and comprehensively than without the game. The case study at the end of this chapter supports this thesis.

Even though games can be powerful in education, they do not have to be complex in nature. In arenas outside of education, the most successful games are ones with straightforward objectives and few rules. We might describe them as simple. Take, for example, the popular video game Super Mario Brothers. The objective of the game is to maneuver the character of Mario around animated, two-dimensional obstacles, so that Mario can get from point A to point B. It is simple. No matter what platform Super Mario Brothers has been produced for over many years—including Nintendo's Entertainment System in 1985 and Nintendo's Wii in 2010—the game is and continues to be popular among hardcore gamers and casual gamers alike, because of its uncomplicated gameplay. And then, there are classic games like Tic-Tac-Toe and Solitaire. No matter the format of these games, tactile or electronic, they satisfy players universally with their simple objectives and streamlined gameplay.

I learned firsthand, in the following case study, that teaching with original games is beneficial to students' educational experiences. My objective in this case study was simple. I wanted to involve students in the process of choosing partners for a team project for class. The process of choosing partners is traditionally straightforward and not very inventive. Here are some of the typical methods and criteria for creating teams that instructors historically have used:

1) Students who sit next to each other in the classroom.
2) Students who possess different skill sets that, in terms of expertise, complement one another.
3) Students of certain ages.
4) Students who live in proximity to another, so work can be easily accomplished outside of class.
5) Students who own different technologies, software, and/or equipment.
6) Students with varying experiences—technological, social, or educational.

The list goes on.

But what if students selected their own teammates? What if they selected their teammates by playing an original game? And, what if, while playing this game, students learned more about each other and the subject matter being taught? If all these hypotheticals were possible, learning could be interactive, informative, facile, and engaging. Students would experience learning as enjoyable, and playing games in education would be productive and relevant. Students might be so engaged that they would crave more and more information. They might want education to go on and on. Then, there would be no arguing about the usefulness of games in education.

So, I set out to see if this were possible, if I could engage students in a game that met the objectives of 1) creating teams and 2) teaching students new material. I would do this by creating an original game.

Case Study No. 6: Learning by Playing Games

Parameters for Devising an Original Game for the Classroom

In order to devise an appropriate game for this case study, I needed to consider a number of both practical and pedagogic issues:

1) The game setting—that is, the environment in which students would play the game—would be drawn either in a graphic design software like Adobe Illustrator on a computer station or by hand with dry-erase markers on a white board, as these were the options available in the class lab. If the computer were used, the digital image would be projected on a screen. Whichever option was employed, projected or drawn, the image would need to be large enough so all students could comfortably view it.
2) There would be no physical game pieces—such as tokens, boards, spinners, and so forth—used for gameplay.
3) The duration of the game would be relatively short, as the objective was to use the game to create teams of two students, for the purpose of a class assignment. It would be counterproductive to fill an entire class time playing the game, because students would need ample time during this one class meeting to begin their team assignments. The team assignments would then continue outside of class and be due the next class meeting.
4) The game concept would be unique, original. If the game concept were derivative or banal, no students, whether experienced in gameplay or not, would be fully engaged. And, engagement was the ultimate goal.
5) The game challenges to be met would be interesting but not difficult. This would create optimal game balancing for the class of under-

graduate students with mixed game-playing expertise. Game balance is a measure of fairness in gameplay. There is more information about this later in the documentation of this case study.

6) By playing the game, students would learn something new about certain course topics, as well as about other students' knowledge of these topics. Learning what and how much other students knew about course topics would guide students in choosing appropriate team members.
7) The knowledge gained by playing the game would aid students in completing the class assignment that would follow.
8) In order to create immediate student interest in the game, the objective of the game—choosing teammates for a class assignment—would be reflected in the game's title: *Choosers and Choosees.*

Educational Game for Team Selection and Student Engagement

Overview

Game Title:	*Choosers and Choosees*
Game Objective:	Students choose teammates for a class project, while learning about the course's subject material
Game Mechanics:	Number selection; player elimination; participation in question-and-answer rounds; project team creation
Game Structure:	Gameplay is divided into four parts: Part I. Choosing Squares Part II. Organizing Groups Part III. Forming Project Teams Part IV. Creating Team Projects

Number of Players

The game works best with a minimum of six students and a maximum of 20 students, and always an even number of students, because the objective of playing the game is to create two-person teams. Solely for the purposes of illustration, the description of gameplay following uses eight students.

Description of Gameplay/Rules of the Game

Part I: Choosing Squares

1) At the beginning of the game *Choosers and Choosees*, each student is assigned a number, one through the total number of students participating in the game. In this case, there were eight students, so students were assigned numbers 1–8.
2) The monitor (the instructor) creates or draws the same number of squares as students, in a graphic design program or on a white board, and presents the squares for all students to see—on a projection screen if using a computer station or other similar display. The squares must be configured in a circular pattern.
3) The monitor numbers the squares by writing numbers in the middle of each square. There will be the same number of squares as students. (See Diagram No. 1. All diagrams appear at the end of this chapter.)
4) The monitor draws a triangle in the top left of every square. (See Diagram No. 2.)
5) Students, in the order of the numbers assigned to them, choose a square. They may choose any square, whether it is the same or different than their assigned number.
6) The monitor writes the number of the student who chooses the square in the triangle. Once a student chooses a square, another student cannot choose it. Remaining students must choose remaining squares, until all the squares have been chosen and numbered.
7) At the end of this part of the game, each square will be numbered, and it will have a triangle in its upper left corner with a number of a student in this triangle. These students are called *Choosers*. (See Diagram No. 3.)

Part II: Organizing Groups

1) The monitor draws a circle in the bottom right corner of each square. (See Diagram No. 4.)
2) To organize students into teams, each student counts the number of squares to the right of *their square*. Their square is the one on which

their assigned number appears in the upper left triangle—that is, the square they chose in Part I. The number students count is that which is written on the middle of their square, not the number they were assigned at the beginning of the game. NOTE: The distinction between *their number* and the number students count is important.

3) The monitor writes the number of the student who lands on the square, after counting squares clockwise, in the circle in the bottom right of the square. (See Diagram No. 5.)
4) More than one student number may appear in the circle, because more than one student may land on that square. These students are called *Choosees*.
5) After all students have counted clockwise and landed on a square, the monitor eliminates any numbered square that no student landed on, meaning no student's number appears in the circle, by drawing an "X" through the square.
6) At the end of this part of the game, certain squares have an original *Chooser*, whose number appears in the triangle, as well as one or more *Choosees*, whose number(s) appears in the circle. And, certain squares have been eliminated from gameplay. (See Diagram No. 6.)

Part III: Forming Project Teams

1) Teams consist of two people: one *Chooser* and one *Choosee*.
2) The first teams declared are *auto teams*. Auto teams are formed when there is just one *Chooser* and one *Choosee* in a square, at the end of Part I. This is unusual and typically occurs only when the game is played with an odd number of students, which is not recommended.
3) The *standard* way teams are created is by elimination, followed by a question-and-answer session. Like all processes in the game, elimination begins with the first square drawn (numbered 1) and goes clockwise.
4) The *Chooser* and *Choosee* cannot be the same number on a single square. Where the number is the same for both, the *Choosee* of that number (in the circle) is eliminated. This may result in one or more additional *project teams*. (See Diagram Nos. 7 and 8.)
5) Additional *project teams* may also result from the previous elimination, because every student can be on only one team. Therefore, the

monitor will need to look at the remaining squares, in a clockwise fashion, to make sure that any student, *Chooser* or *Choosee*, already on a team, is eliminated and not coupled with any other student. (See Diagram No. 9.)

6) At this point in the elimination process, any student who, across the board, is both a *Chooser* (their number appears in any triangle) and a *Choosee* (their number appears in any circle), becomes a *Chooser* only. Their number remains in the appropriate triangle but is deleted from all circles. (See Diagram No. 10.)
7) If at this point, any two *Choosees* are left together in a single circle, and there is no *Chooser* in the square, then the *Choosees* are put in Limbo. Players in Limbo are not eliminated from gameplay. They are just not actively involved in gameplay—temporarily.
8) If all numbers have not been paired and teams formed according to the process described earlier, and a *Chooser* still has more than one *Choosee* to pick from in his or her square, then a question-and-answer round occurs. (See Diagram No. 11.)
9) In order to make informed choices about teammates, the *Chooser* may ask each *Choosee* three questions that pertain to the class project that individual teams will work on.

   For example, if the project is a website for a not-for-profit arts organization, the *Chooser* may ask the following:

   a. How would you best represent the mission of the arts organization with a new website?
   b. How do you envision designing the user interface so that visual balance tips more toward images and less toward text?
   c. How does a hexadecimal that represents web-safe colors differ from one that represents millions of colors? Please provide examples.

10) After every one of the remaining *Choosees* has answered the same set of three questions, the *Chooser* picks the one *Choosee* whom he or she believes will be the best teammate. (See Diagram No. 12.)
11) The *Choosees* not chosen after the question-and-answer round are put in Limbo by the monitor.

12) The final teams are created with the students in Limbo. If there are more than two students in Limbo, the monitor puts them in teams of two, in a way that will best serve the project. For example, students with complementary skill sets and different levels of expertise are paired together. (See Diagram No. 13.)
13) Gameplay concludes with the formation of all project teams. (See Diagram No. 14.)

Part IV: Creating Team Projects

When the game is over and teams are selected, the instructor (monitor) gives students the class assignment, and they begin working on their team projects, richer for the experience of selecting their own team members, as well as learning more about the course topics during the question-and-answer rounds in the game.

Observations and Conclusions

In this case study, I played the game *Choosers and Choosees* three times, once for each of three of my courses—at beginning, middle, and advanced levels. There were eight students in the first (beginning) group, 12 in the second (middle), and 16 in the third (advanced). The results were similar for all groups, as follows:

1) By playing the game *Chooser and Choosees*, teams for a class project were successfully formed.
2) During the question-and-answer phase (Part III: Forming Project Teams), student *Choosers* asked appropriate questions and student *Choosees* provided sufficient answers, demonstrating that students understood the course material.
3) Students' attentive behavior throughout the process showed that they were fully engaged: playing the game and choosing teammates for the class project.

This last result is important, very important. During this case study, while students played the game, I observed a level of engagement much higher than I usually see in the classroom. Students were *genuinely interested* in how the game unfolded from beginning to end, how they were involved in the process of choosing their teammates, and what the results of the game would be in terms of the formation of all teams for the class project. In addition, there was an overall positive spirit in the class, which I attribute to team morale.

During gameplay, there was no multitasking, intentional or unintentional, by students. Students in the case study chose not to engage with mobile devices—for texting, browsing, or any extraneous activity. Instead, students were focused on and involved with the game, the class, almost completely. This kind of participation recalls the discussion of *participatory pedagogy* earlier in the fifth chapter of this book:

> When the *interact* component of interactive learning is taken to the next level, students are more involved in the educational process, and their participation is consequential. With new media technologies, students become partners in shaping pedagogy.

This case study showed that games in education, when designed to serve pedagogy, work. And, in terms of advancing student engagement with course material, games in the classroom are powerful.

Diagrams for *Choosers and Choosees*

Diagram No. 1

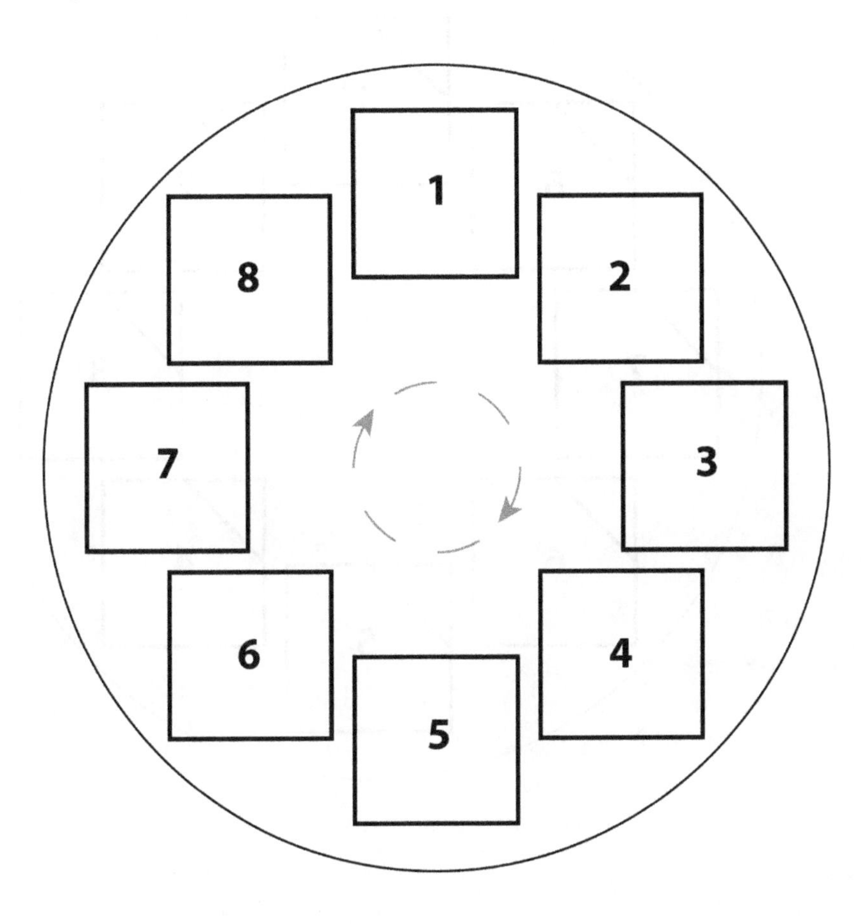

Diagram No. 2

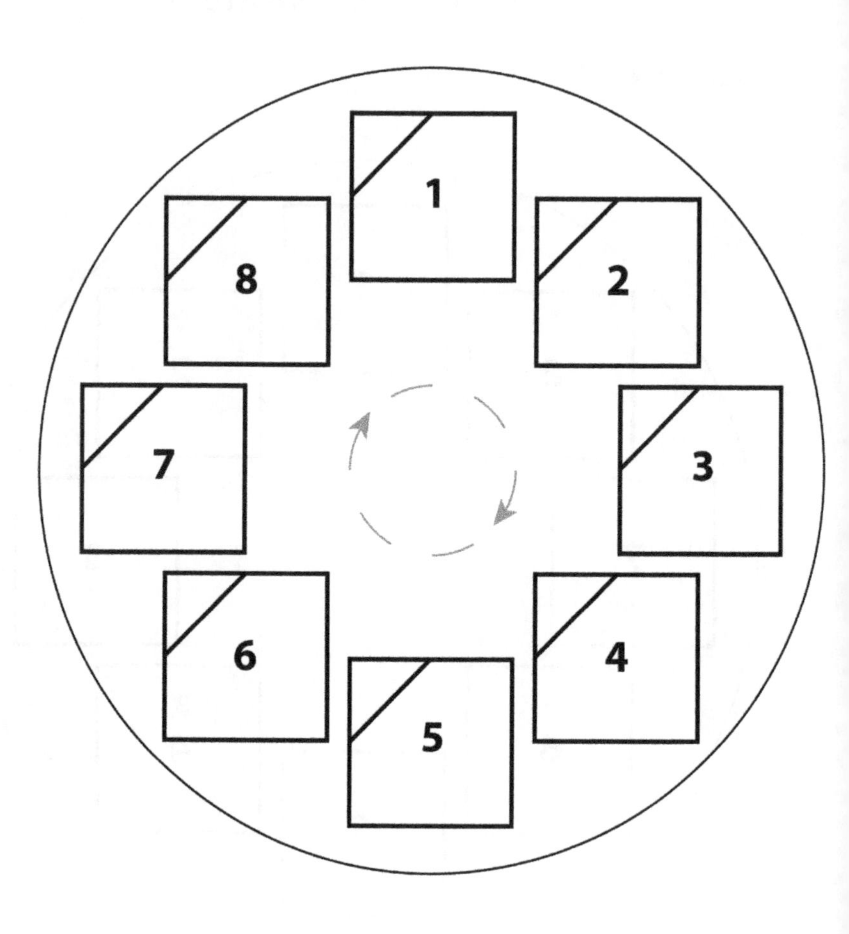

Diagram No. 3

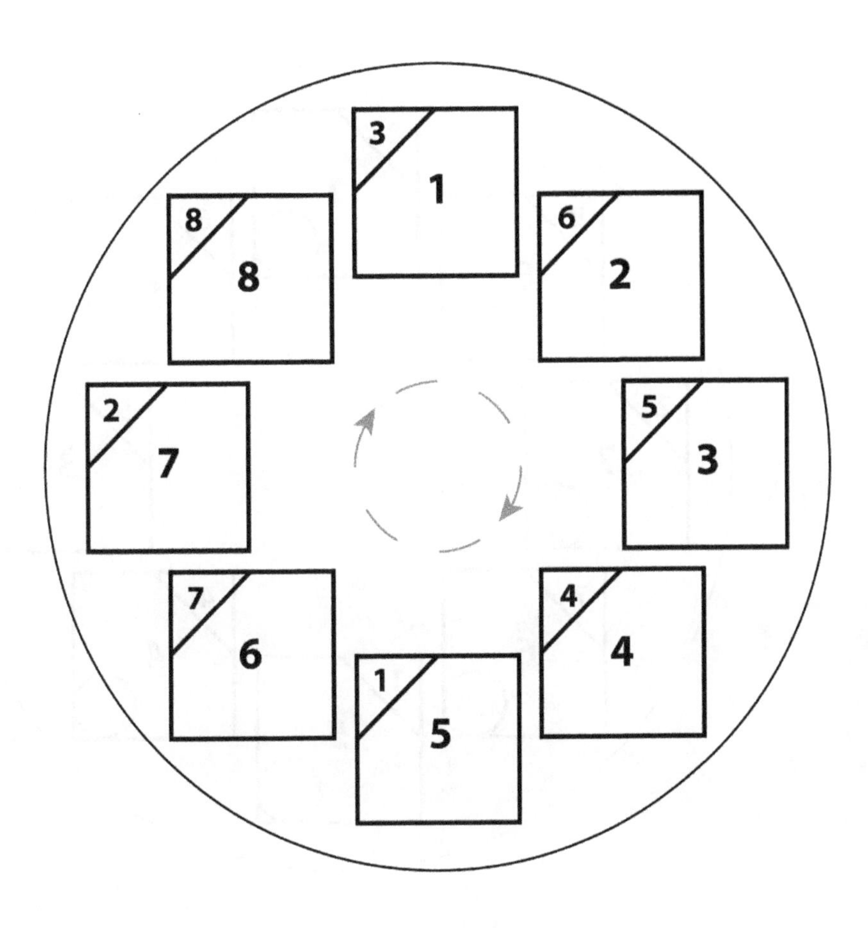

Diagram No. 4

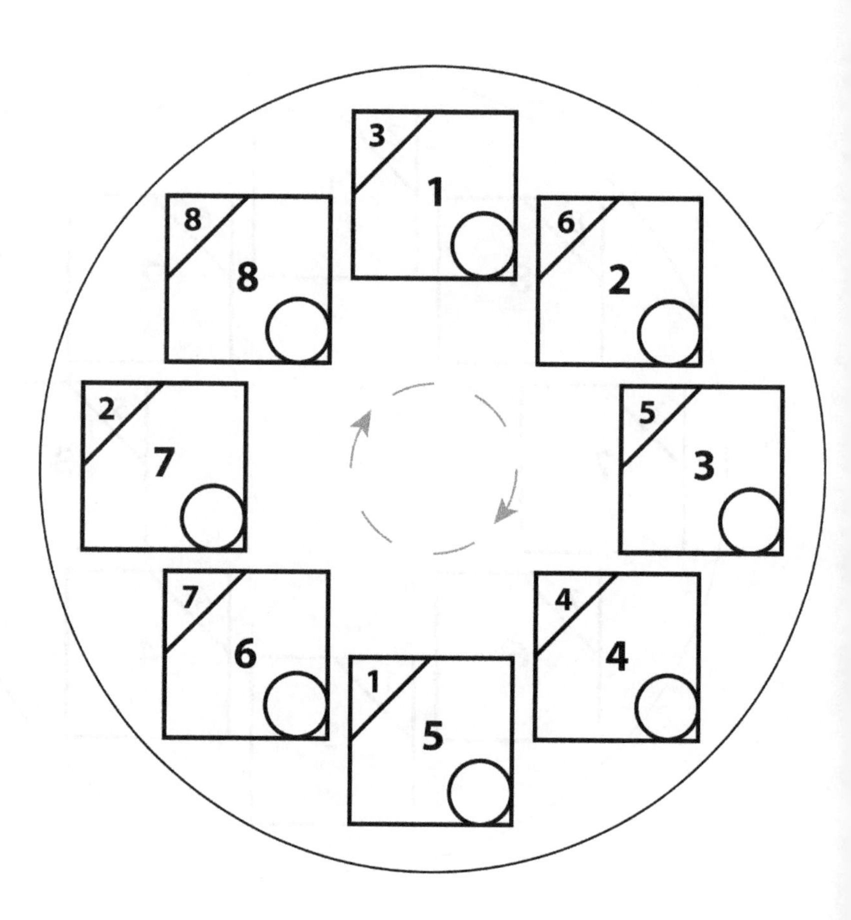

Diagram No. 5

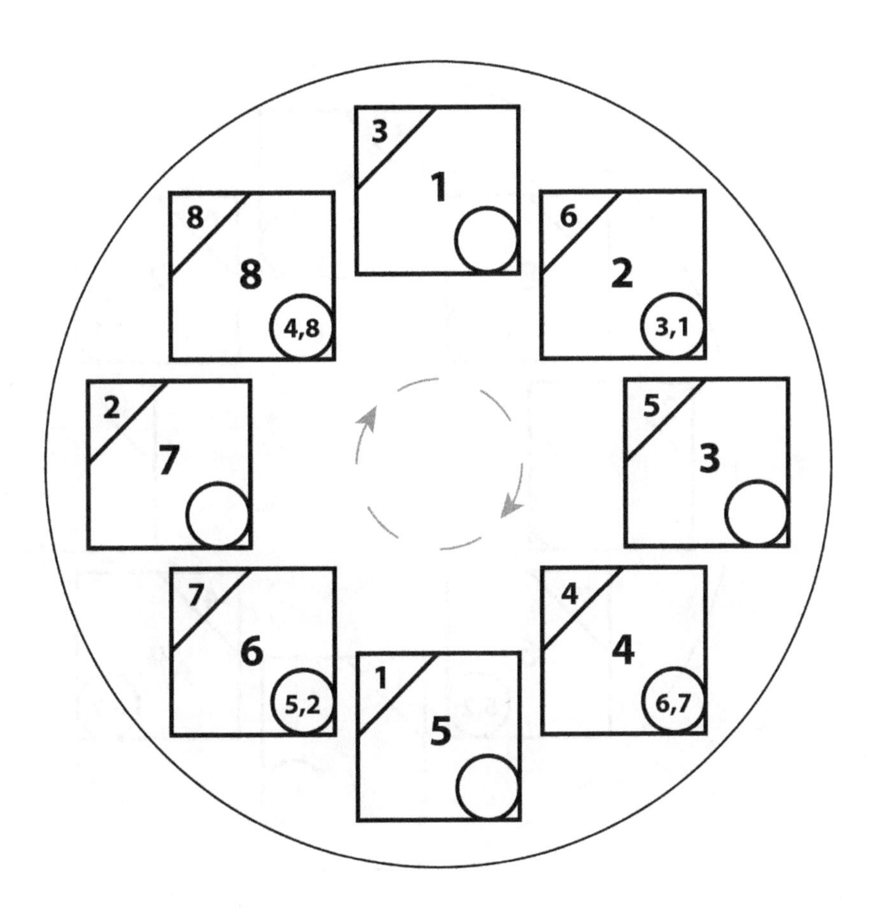

Diagram No. 6

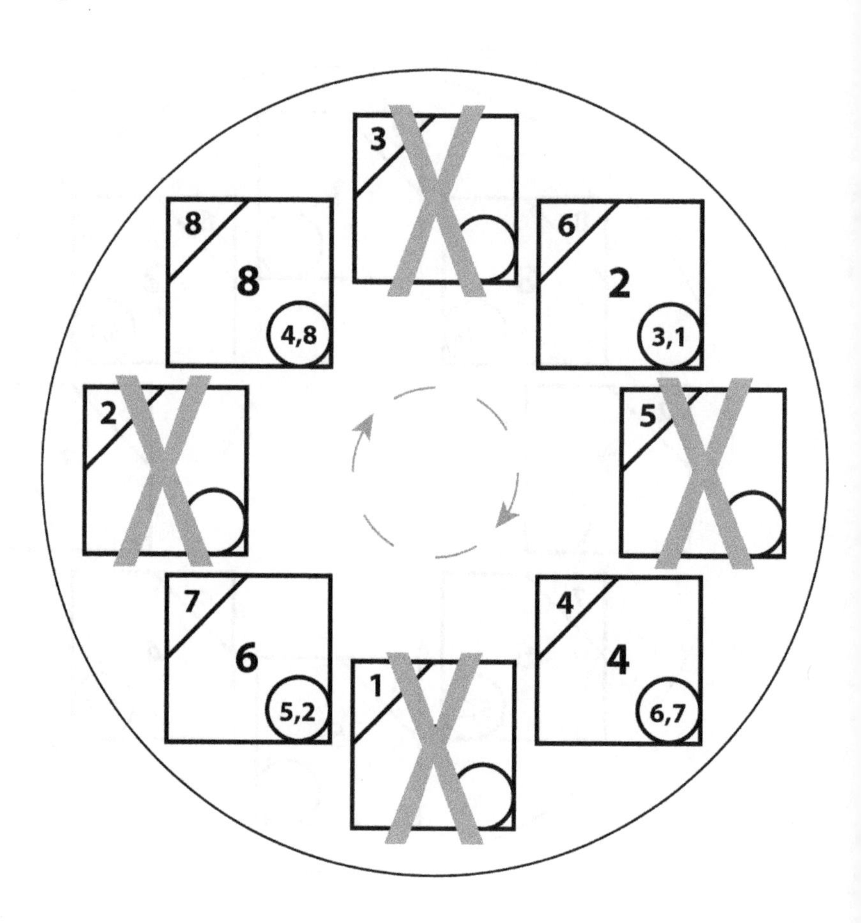

Diagram No. 7

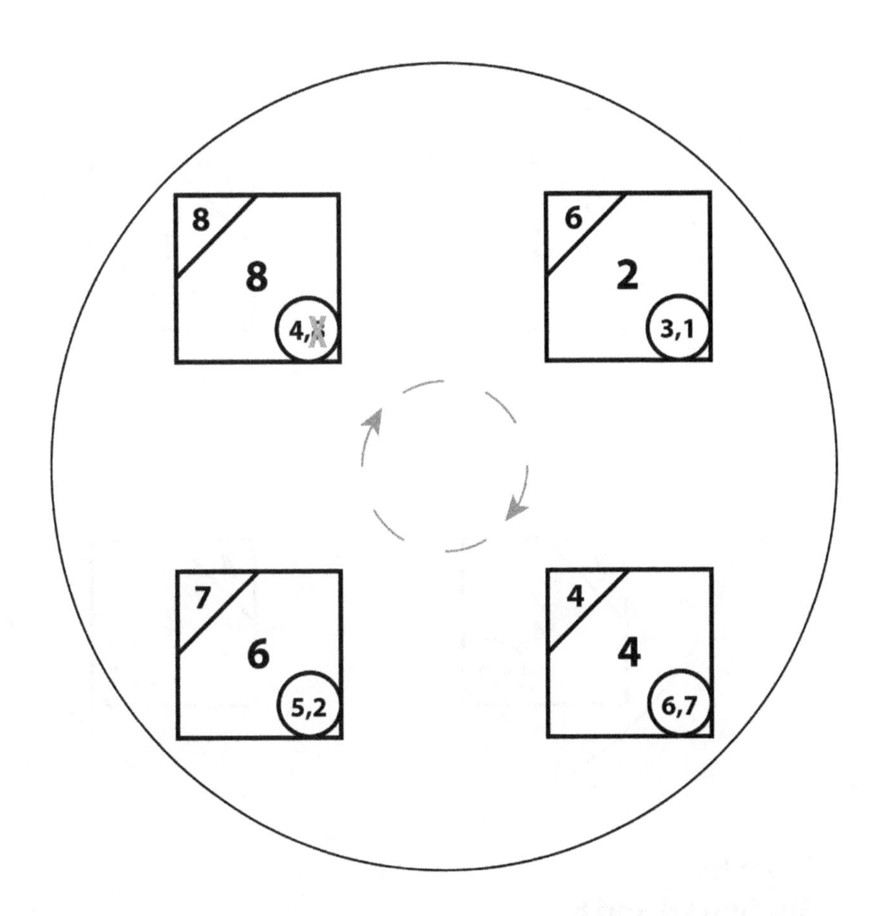

Diagram No. 8

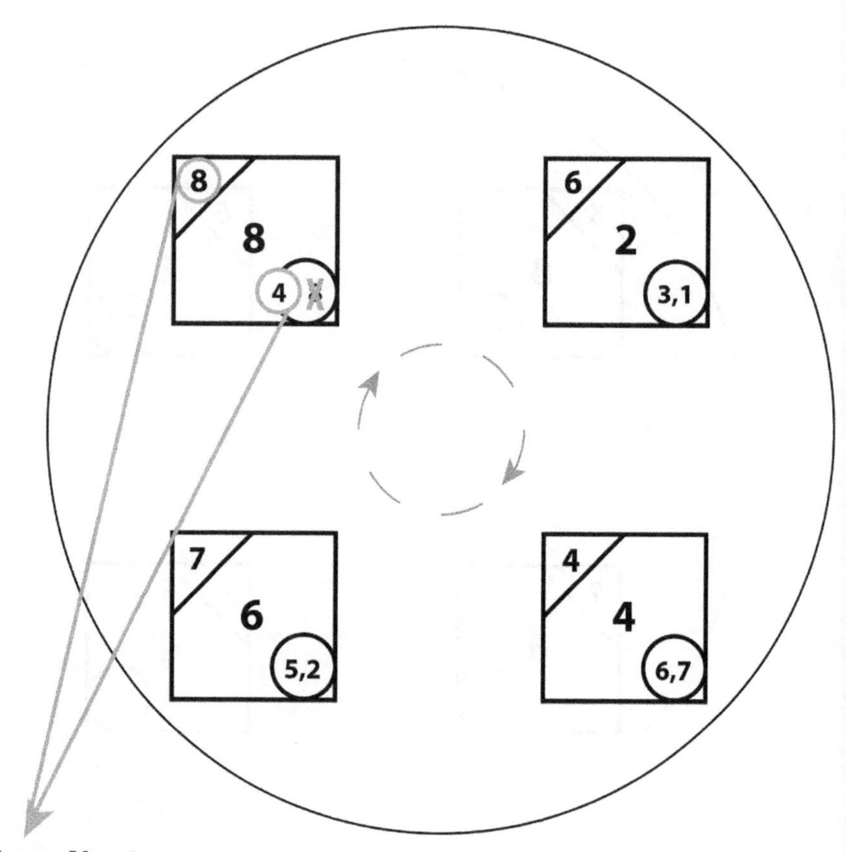

Team No. 1
Students 4 and 8

Diagram No. 9

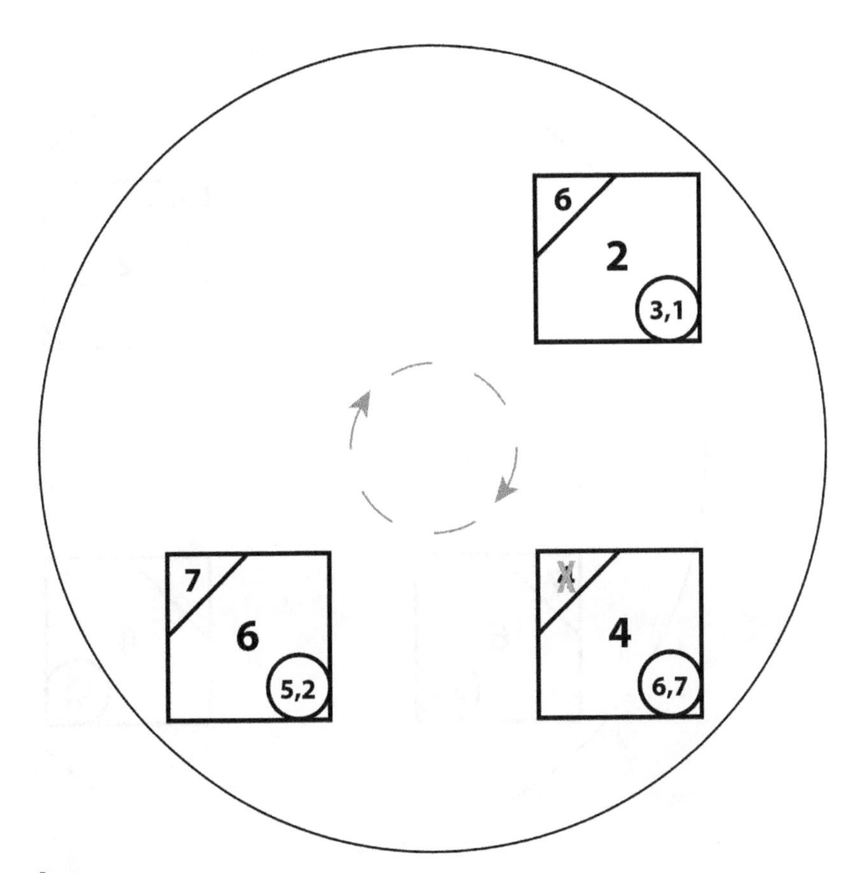

Student 4
eliminated as a Chooser

Diagram No. 10

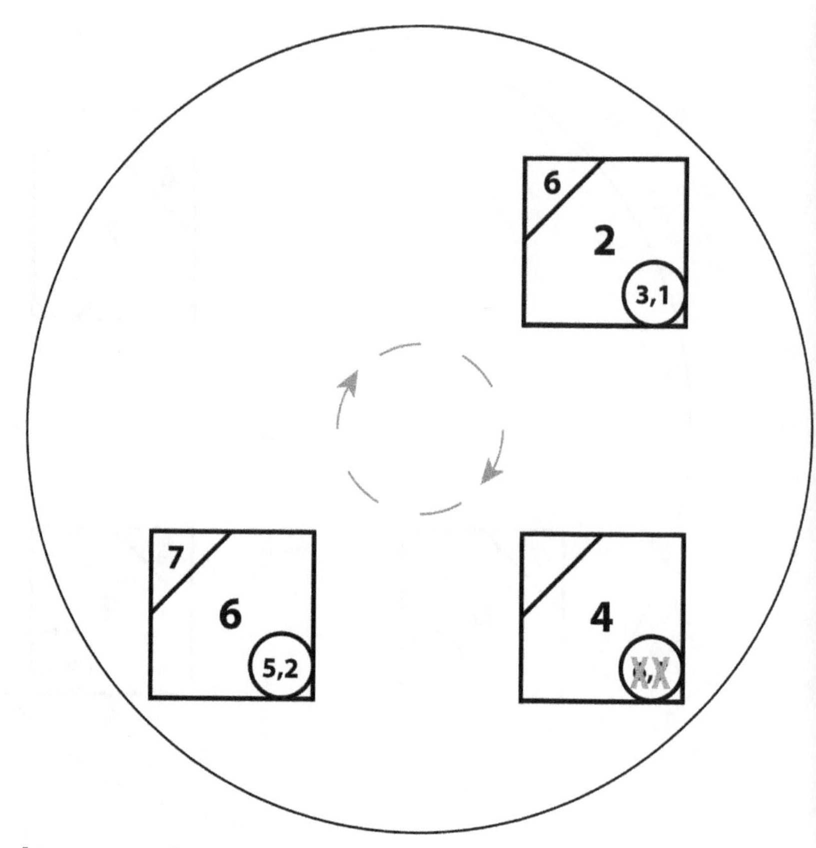

**Students 6 and 7
become Choosers only**

Diagram No. 11

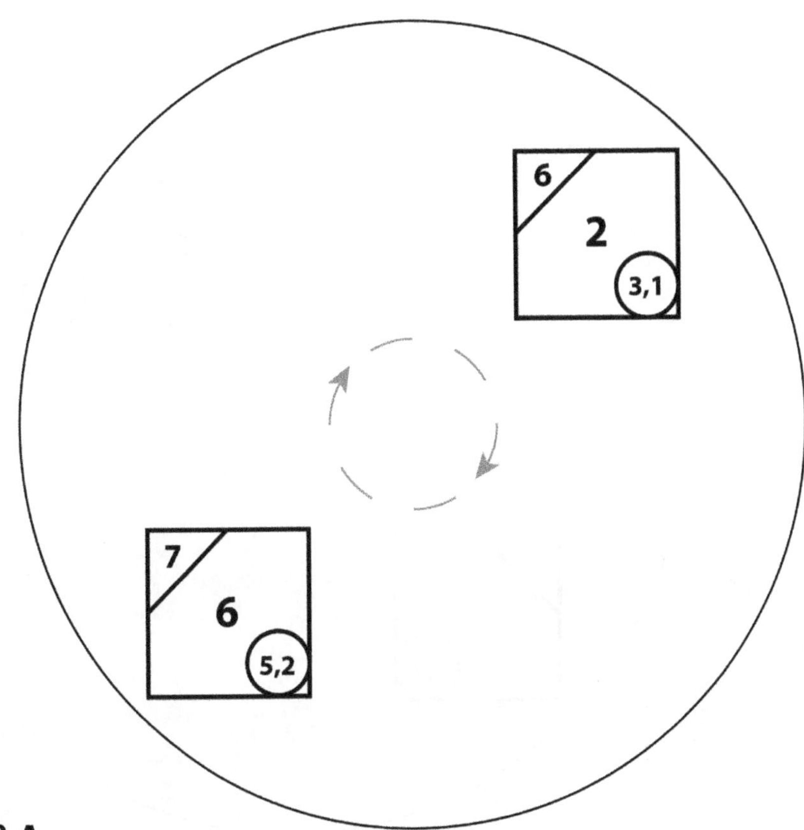

Q & A:
Chooser 6 questions Choosees 3 and 1
Chooser 7 questions Choosees 5 and 2

Diagram No. 12

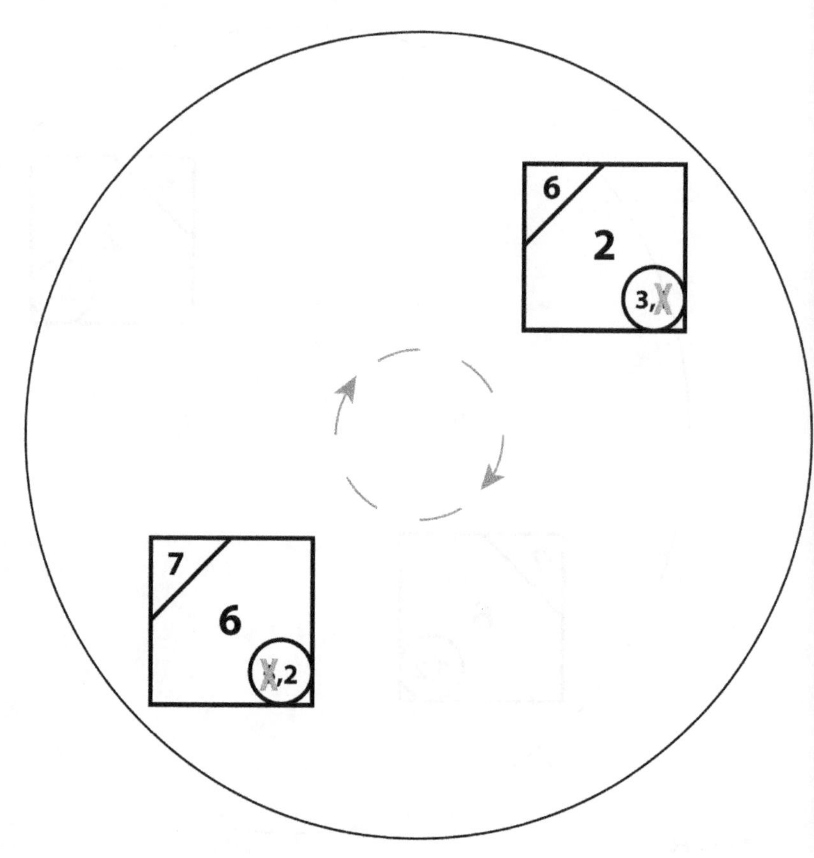

Team No. 2: Students 6 and 3
Team No. 3: Students 7 and 2

Diagram No. 13

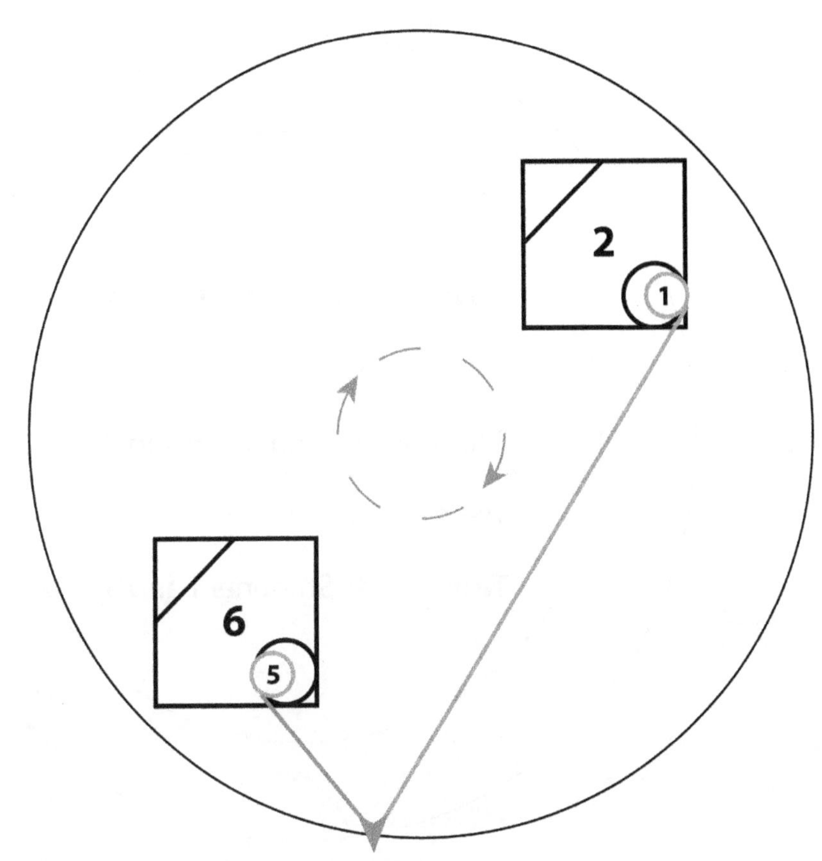

Students 1 and 5 are put in Limbo, become Team No. 4

Diagram No. 14

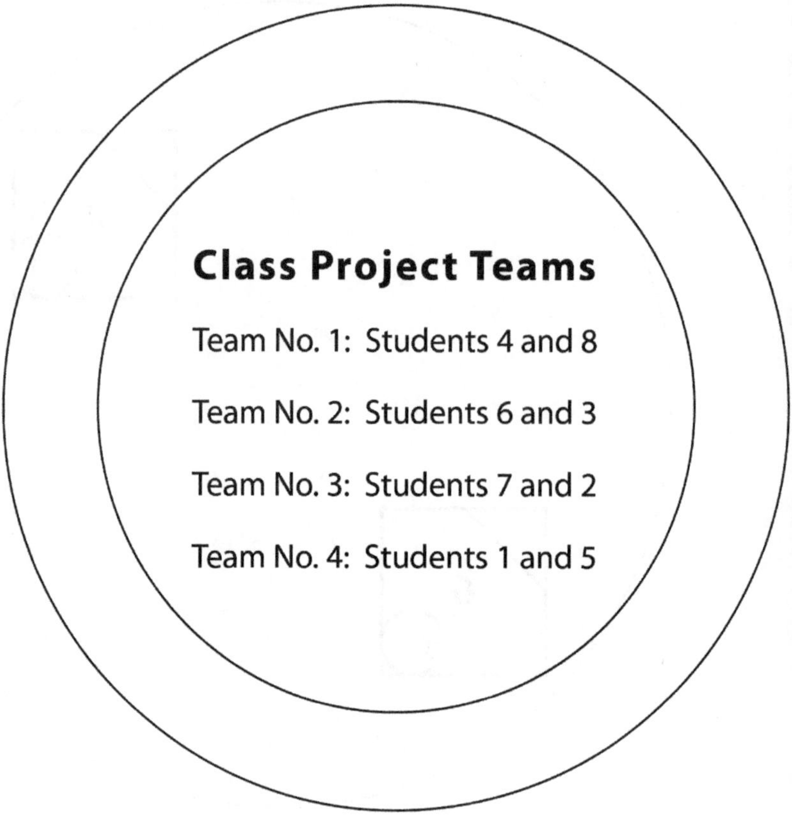

Class Project Teams

Team No. 1: Students 4 and 8

Team No. 2: Students 6 and 3

Team No. 3: Students 7 and 2

Team No. 4: Students 1 and 5

15

THE AMORPHOUS CLOUD

The cloud is a big, ill-defined, useful, fuzzy, expansive collection of computers strung across the Internet, which ostensibly allows users to store data at any time, of any size, and in various digital formats, and to access that data whenever they want. The cloud is as much theoretical as it is practical. It is theoretical in that users believe they can store as much data as they want without limitation. It is practical in that it is a collection of interconnected computers and computer systems that are real and finite in their ability to store and transfer data. It is theoretical in that users believe that big, commercial service providers like Google give storage for free. It is practical in that there is a definite cost associated with using the cloud, but it may not be monetary. Like an atmospheric cloud is amorphous, free-floating, and without boundaries, so is the notion of the cloud in the world of computers and new media.

The cloud could be described as a virtualized data center hosted on servers around the globe, where pools of data are stored and accessed by the data's owner or whoever else is connected to the cloud and can gain access to the data. The abstract nature and vastness of this pool of data can make it appealing in higher education, because the cloud stores a bounty of information available via the web. Although we educators like to believe that a repository of data is a good thing in education, when it comes to new media and technology that serve pedagogy, more is not always better. When aggregating, storing, and serving up virtualized data pools across networks of computers worldwide for educational purposes, there should be checks and balances put in place to protect that data, as well as filters that allow instructors and students to separate data that are appropriate for learning from data that are not—digital noise. In most cases, security measures and filters are missing in the cloud.

It is easy for all of us—new media educators, students, and administrators—to believe in the limitlessness of the Internet and vastness of the cloud. When we surf webpages, participate in social networks, or upload digital files to database-driven websites, we can work with files that are very large in size and for very long stretches of time. Though some file size limitations exist, there are often upgrades or work-around maneuvers that let you bypass them. For example, on the storage and courier website YouSendIt, users can send files of a certain size free of charge, then upgrade by paying a fee to send larger files. With social video sites like YouTube and social audio sites like SoundCloud, users can upload many files free of charge and frequently. This seems like a great deal: abundant storage and facile distribution of digital content to a global audience at no, or little, cost.

But there are several important things to remember:

1) Nothing is free. These cloud websites make money from advertising on their sites and by selling users' demographic data—captured when online users fill out profiles—to marketing concerns and other for-profit entities.
2) Someone or some entity owns the data. We must remember that individuals, institutions, organizations, or businesses created, authored, and/or produced the content that has been uploaded to the cloud. The owners of this content have ownership rights—such as copyright, trademark, and so forth—that should be respected, but they are not always.
3) Someone or some entity owns the servers. The computers on the Internet that make up the cloud—whether accessed via wireless technology, satellite, cable, or telephone line—are owned by individuals,

institutions, organizations, or businesses. The owners of the servers and the owners of the content are not likely one and the same.

Ownership issues are real in the world of the cloud. Creators of original content own the data files that are uploaded to the cloud, either for free or pay. These creators could be writers, digital artists, multimedia producers, game developers, interaction designers, and so forth, and they could work in either the commercial or noncommercial arena. It seems that the creators would want to protect the data. Yet, with a remarkable false sense of security, we upload file after file after file to someone else's server, rarely questioning where our data is going, where our files are living, who has access to our content, or what those people may be doing with it. Similarly, we do not inquire about what security measures are in place to thwart hackers and other cyber criminals from stealing our data.

Why is this false sense of security so prevalent? Why do we not ask more questions, make more demands, have more control? Why are we okay with this Faustian bargain? The answers to these questions lie in the fact that too often technology companies persuade us with their marketing pitches, and we buy into the hype. Or, perhaps we just want something for free!

For example, take Apple's iCloud service. On the Apple website, the service is touted as follows:

> This is the cloud the way it should be: automatic and effortless. iCloud is seamlessly integrated into your apps, so you can access your content on all your devices. And it's free.[33]

Free. I do not see how it is free. We need to purchase an expensive Apple device, such as an iPhone, iPod Touch, or iPad, to use iCloud. On top of that, the iCloud service is proprietary. If we want to add a Sony tablet computer or an Android smartphone to our collection of devices, we will have to back up the data on those devices to a part of the cloud not owned and controlled by Apple, not to iCloud. That begs the question: With iCloud, does Apple own part of *the* cloud, or does it just own *i*Cloud? Is the cloud different than iCloud, or are they one and the same? Then, there is all that privately owned user data that is "effortlessly" transported to iCloud almost

---

[33] http://www.apple.com/icloud/

magically, with the press of an icon on a device. If Apple owns iCloud, whether or not it is part of the cloud, and we content creators can use iCloud for *free*, does that mean Apple owns or at least has access to our content on iCloud? It is confusing.

At the time this book is being written, Sony has created its own cloud service called PlayMemories Online. It seems like an overwrought name for data storage on a server. But nevertheless, in one warm, fuzzy, accessible, and free place online, users can upload their digital memories in the form of videos and photos that they have captured on compatible devices, such as Sony televisions, computers, tablets, and so forth. Sony, too, chooses to redefine *free*. Purchasing the compatible equipment necessary to use PlayMemories Online costs money, sometimes lots of it.

Faust, with all his bargains, seems to be doing pretty well in the cloud, because now he has access to and perhaps ownership of our photos, music, documents, e-books, e-mails, contact information, and applications. An Apple promotional video for iCloud boasts that, "you always have the things you want exactly where you want them."[34] Convenience is one thing, but personally, I would like a guarantee from iCloud and all cloud services, for that matter, that my content is secure, that it cannot be tampered with or accessed by other people, companies, organizations, or entities around the globe without my permission. With the prevalence of computer network hacking and, beyond that, the potential of cyber terrorism, it may be a long time until Apple or any company can make that kind of guarantee.

In addition to ownership concerns with the cloud, there are privacy issues that need to be addressed. For years, the web was a world of anonymity for those who wanted to surf online content, post comments on discussion boards and in blogs, investigate content on websites, and generally explore a binary world of scintillating tidbits and curious information. If we did not want to divulge our true identities, we did not have to. Over the years, there has been much discourse and debate about the pros and cons of anonymity on the World Wide Web. Questions that continue to be raised include the following:

1) Should you divulge personal information, such as name, phone number, and street addresses, online?

---

[34] http://www.apple.com/icloud/what-is.html

2) If you are curious about a particular niche community online and would like to receive information from it, but feel there may be personal or professional repercussions from being associated with it, should you join the community?
3) Should you provide your real identity when participating in online games or visiting virtual worlds, where the experiences are meant to be about fantasy?
4) Should you be yourself or an enhanced version of yourself on job and dating websites, if you feel it is to your advantage to put your best foot forward?

These are just some of the questions we ask ourselves and some of the dilemmas we aim to resolve when going online. Answering these questions is important. Assessing what information to reveal to the global online community, protecting our identity, and keeping tabs on what personal information becomes public is even more important. How can we possibly do this in the amorphous, unprotected cloud? I want security as well as privacy protections, but I wonder if these are really possible with the cloud. These are consequential issues when using the cloud in education.

As expansive as we view the Internet to be, or as limitless as we perceive the repository of data in the cloud to be, we must always pause and remember that the Internet is a computer network. It is physical, not theoretical. For students, it is necessary that we educators reconcile the theoretical with the physical, when including the cloud in our lesson plans. It might be current, even hip, to lecture students about the mysterious, endless repository of data that they can access with a computer, laptop, smartphone, or other device, and that their education is richer because they can learn from this repository of data as well as contribute to it. But in our teaching, we need to decide how we can integrate something that exists *in the clouds* with something that requires *real and dedicated* access via an Internet connection, usually for a fee, not for free.

The first thing we educators must establish for our students is that the cloud cannot be a source of information and a myth at the same time. Educators who support the myth of the cloud do so by assuming that the technology that supports the cloud will continue to develop, so that future scalability of the cloud will be guaranteed. We do not know that future technology will support the kind of scalability and growth that allow everyone to store and access as much data as they want all the time. People who believe in the myth hope that it will. They presume that it will. Because technological development has occurred so rapidly and consistently over the past couple of decades, it is easy to presume that such development will continue into the future. This is similar to the *presumption of connectedness* discussed earlier in this book—when referring to students' expectation that we are available via electronic communication constantly. Just as we cannot presume a state of complete connectedness, we cannot presume continual development of cloud technologies.

As I write this book, the global economy is in a recession. Certain national economies are crumbling. This will likely have an effect on investment in new technologies and the future development of the cloud in the near future. More than that, people's attitudes shift. We cannot presume that society will *want* to remain so connected electronically to each other and to the cloud, with or without technological developments. There may come a time when we want to take our data and store it in a smaller, private, more secure environment, not one that is accessible by vast numbers of people and vulnerable to Internet security breaches.

A look at trends in software development over the years illustrates our desire as content creators to possess and control our own data. It was not so long ago that technology enterprises were keen on application servers. Generally, application servers provide a software environment on a client-server that allows users to access and use software programs stored on this server, even from a remote location. Users do not have to download a separately purchased software program with its own license onto their individual computer or work station. Theoretically, this is wonderfully efficient: connect to the programs you need, when you need them. However, these connections, made over a computer network or via the Internet, can prove unreliable. Networks fail. Internet connections can be spotty. In addition, working with large files, such as video and multimedia files, via remote, server-side software, can be difficult, and files can become unstable. Because owning outright the software programs, protecting data files, and working in a stable computer environment are all important to users, application servers fell out of fashion.

There is an attempt by Adobe to resurrect the concept of shared applications, albeit with a twist, with Adobe Creative Cloud. According to Adobe's website, "Adobe Creative Cloud is an ongoing membership that lets you download and install any of the new Adobe Creative Suite desktop software."[35] The user downloads the applications onto a local computer, which is different than an application server, where the software programs would be stored on remote computers. However, with Adobe Creative Cloud, access to the software tools is monitored by cloud technology. Users, depending on the payment plan they purchase, can access and use the applications for set periods of time. Certain plans allow users to store their data in Adobe's Creative Cloud as well. This mix of application-server concept with cloud technology is an interesting option for users who want or need software but do not want to purchase it outright. Time will tell if this approach is successful for people who want software and storage at a relatively affordable price.

Along with systems like Adobe Creative Cloud, there is the growing popularity of applications for mobile devices, which is another way software usage is trending back to a centrally controlled system of distribution. But something is different with the mobile application. We purchase an application, download it to our device, and then connect wirelessly to a server system for certain data, if and when that data is needed. There is no external monitoring of our use of the application, ostensibly. Of course, there may be tracking of our use of the application and location (via GPS coordinates) by the manufacturer of the application for its own internal purposes. But this tracking does not usually determine how and when we use the application, and it does not necessarily affect the application's core functionality. So, there is outright ownership of the application on some level.

In addition to the issue of ownership, there is the issue of privacy to consider. With the rise of user involvement online and an overall increase in user engagement with web-based content—including the public's participation in blogs and virtual communities—there are privacy issues that should be addressed more fully than educators, government regulators, Internet entrepreneurs, and technologists are addressing them at present. The issue of privacy with new media is especially important with cloud storage. We are depositing more and more data in the cloud, and more than ever before, users around the world want to comment on, change, and exploit our data. When appropriate, we must place restrictions on access to the data in the cloud, in order to protect the privacy of all individuals associated with the data—whether those individuals are owners, subjects, authors, or just people mentioned in the data.

---

[35] http://www.adobe.com/products/creativecloud.html

There cannot be a full discussion of privacy, personal data, and the cloud without mentioning social networks. While we are still in the infancy of the development of these platforms—in terms of their long-term impact on society—there are no other technology systems currently available that connect everyday people to such vast pools of data and allow users to access and share data so frequently and copiously like social networks. There are few regulations in place that control how social networks acquire, manipulate, and serve up other people's data. Though some governments around the globe have stepped in to monitor potential violations of citizens' privacy by social media companies, for the most part, social networks are controlling users' data *their* way. Let us keep an eye on the future of these systems and work to create rules and regulations that will direct all social networks to control data *our* way, so that privacy is protected and ownership of data is respected.

The topic of data and privacy, as it pertains to online platforms and social media, is one that has been written about a lot. There are entire books devoted to it. In this book, which is focused primary on new media in education, I bring up the topic, not to investigate it in detail but to illustrate the pervasiveness and power of data online and to call attention to the need for official oversight of the use of other people's data. This is not a call for censorship or the implementation of draconian monitoring of data, at all. This is simply a way to say, hey, let us all work to protect privacy and respect ownership of intellectual property. When using the cloud in instruction, we educators have an obligation to communicate all of this to students.

Recently, I was an invited presenter at a symposium on the topic of technology in education. I presented alongside an accomplished medical researcher, whose work was quite impressive. The audience was invited to ask questions after each presentation. One member of the audience asked the researcher how she envisioned future students learning from her research. The researcher responded, "I'll put it in the cloud." It sounded modern, progressive, and even benevolent. But I am not sure she understood the potential ramifications of what she was saying.

If she put all her research in the cloud, she would put years of work done by her and her team of researchers in a place that made the information available to the world at no cost. Medical research is expensive. It receives funding from many sources. These funding sources, which would likely include universities, government agencies, and foundations, may want a say in when, how, to whom, and to what extent the research is made available, especially if these entities were interested in a return on their funding investment. Responsibilities to funders aside, my fundamental point is that in an effort to fit in with the technological elite, or in the heat of a technologically sexy moment, or because we sometimes fail to see the whole picture with regard to the cons of the cloud, we just stick our data in it, without addressing the important issues of ownership. Simply dumping data in the cloud does not make scholars tech-savvy.

Believers of the cloud myth perpetuate it in and out of the classroom, because they presume the cloud is tangible and the cloud will continue to be functional with future advancements in technology. There certainly is room for myth in education, when discussing and teaching topics related to mythology, such as Greek and Roman, not when discussing and teaching topics related to technology. Let us keep myth in the storybooks. Hardware is hard. It exists.

If we want a clear definition of the cloud, as well as an understanding of how it really functions, we would be better served by referring to the cloud simply as a network of computers interconnected via the Internet, because that is what it is. The Internet is real, and it provides meaningful connections, data storage, and access to real information. All of this is valuable, but all of this is finite. Until there is international consensus on what this vast network of storage computers should be called, it is likely we will continue to refer to it grandly and mythically as the cloud.

16

NEW MEDIA'S TRANSFORMATION OF EDUCATION

Web 2.0, social media, websites, wikis, interactive forums, e-learning systems—at some point they are all part of the educational process today, no matter what the discipline. New media has transformed education by seamlessly integrating technology into pedagogy. The results of this transformation are sometimes consequential and innovative, sometimes insignificant and uninspiring, sometimes worthy, sometimes not.

Jean Piaget, the psychologist and philosopher, said, "The principle goal of education in the schools should be creating men and women who are capable of doing new things, not simply repeating what other generations have done."[36] New media technologies provide a means for doing new things—in education, art, community service, and industry. And as long as technologies continue to change, there will always be new things we can do with new media. New media have the potential to enhance the overall quality of an educational experience, even a life experience, as long as educators understand how theory and practice should converge with technology in the classroom and educators put purpose first.

Because convergence is a vital component of our electronic media culture, educators should provide students with a strong conceptual and methodological grasp of issues in the intersecting worlds of emerging

[36] Jean Piaget, "Piaget Rediscovered" in *Journal of Research in Science Teaching*, 2, Eleanor Duckworth (1964): 175.

technologies, new media integration, and content creation. That is my approach. I believe that with an understanding of how content, media, and culture converge to engage users—visually, interactively, or otherwise—students will be better prepared to produce meaningful works in the changing digital landscape. Fusing technology with human experience is how we as a society continue to do new things with technology. We teach this to students and hope they will take their meaningful classroom experiences, with electronic pads, e-readers, tablets, and mobile devices, into their chosen professions. If our teaching with devices and technology is effective, students will do just that and thus expand the digital landscape around them.

In *The New York Times* online article "Inflating the Software Report Card," writers Trip Gabriel and Matt Richtel explain how publishers sell software and educators use this software to teach—while neither party really understands how the tools work or whether the tools are effective in education.

> School officials, confronted with a morass of complicated and, sometimes, conflicting research, often buy products based on personal impressions, marketing hype or faith in technology for its own sake.[37]

My advice here is this: Avoid the hype. With so many technology products out there, educators must assess what tools genuinely meet pedagogic ends before purchasing them, and certainly before integrating the tools into instruction. While there is some room for trial and error after tools have been purchased, no educator wants to spend a significant amount of money on technology that cannot be used effectively in the classroom or that serves no measurable pedagogic end.

Know when to use technology and when not to. Core subjects like mathematics and English have been successfully taught in schools for years, long before software manufacturers produced tools to help do this. Can the click of a computer mouse facilitate the learning of multiplication tables, or do students just have to memorize them? Is the rule *i before e except after c* more applicable if a student learns it on a tablet computer as part of a distance learning course or in person in a bricks-and-mortar classroom? The answer to both of these questions, quite simply, is the same as the general conclusion I have drawn from the six case studies conducted for this book:

---

[37] Gabriel and Richtel, "Inflating the Software Report Card."

Using expensive technology is not necessarily a better way of engaging students; it is a different way.

In the precarious environment of changing technologies and evolutionary pedagogies, a debate does and must continue over the effect of technology on learning. However, I am confident that educators on both sides of the debate will, at some point, recognize the far-reaching potential of new media tools and devices and put them in perspective. An animated, media-rich user interface may stimulate a student in a way that a human voice or hand gesture cannot, but it does not engage a student any more reliably. Both the technological and the traditional are important. They are partners on the journey that is education in the 21st century. The key is to use one *with* the other, not instead of the other, to serve learning outcomes, and at a reasonable cost.

Some educators believe that technology is the panacea for our educational woes and that technology is worth the cost, no matter what the cost, because our students will not succeed academically or professionally without it. I cannot imagine that the teacher of a young Abraham Lincoln, in a one-room schoolhouse in the early 19th century, would have needed an iPad. There was no digital landscape back then. And still, complicated problems got solved. However, it may not be that simple. As mentioned throughout this book, today technology is integrated into our lives socially, academically, and professionally. New media, convergent and otherwise, are part of our consciousness. Now, the digital landscape is everywhere. Whether or not we embrace technology in education, and whether or not we deem it beneficial or detrimental to pedagogy, new media *must* be put into the mix. The times require it.

There is no going back to traditional methodologies alone. The focus moving forward needs to be on creativity and integration: creativity of mind and spirit and integration of tried-and-true approaches with new ones. In education, if new media are a distraction, let us make them an enhancement. We should be innovators, the artists that Seth Godin encourages us to be. And, we should heed the advice of Buckminster Fuller, to establish a new form of education that is "complimentary to the innate faculties and capabilities of life."[38]

Because we live in a democratic society that respects freedom of thought and expression, and we live in a culture that encourages educators to harness the power of that thought and expression, it is indeed possible for us to estab-

[38] Fuller, *Education Automation: Freeing the Scholar to Return to His Studies*, 7.

lish a new form of education, with new media as its foundation. This is education conceived to work in the global digital landscape. This is new education. Because of technology, new education permeates cultures and crosses national borders. New education is key to building and sustaining modern, open-minded, and democratic societies. Therefore, we must educate tomorrow's leaders of industry, government, and the arts with it.

With the ever-changing state of new media, there will always be new challenges to meet using technology in education. How do serious educators of and with new media create focused pedagogy and engaging classroom experiences? The answer is simple: consistently be inconsistent. Change up the tools, technologies, and methodologies that we integrate into teaching, while embracing traditional theories and methodologies when they apply. Experiment when necessary. Focus on the consequential, innovative possibilities of new media. And, do it all with purpose. Make sure the integration of technology in pedagogy has been tried in the classroom, evaluated out of the classroom, and most importantly, that the tools serve curricula. Remember that technology does not educate students, we do. Then, new media will transform education in a significant way. Web 2.0 will produce Education 2.0.

BIBILOGRAPHY

Brooks, David. "The Other Education," published November 26, 2009, http://www.nytimes.com/2009/11/27/opinion/27brooks.html. Accessed April 7, 2011.

Bruner, Jerome. *Acts of Meaning: Four Lectures on Mind and Culture (Jerusalem-Harvard Lectures)*. Cambridge, MA, Harvard University Press, 1990.

Clark, Andy. "Out of Our Brains," published December 12, 2010, http://opinionator.blogs.nytimes.com/2010/12/12/out-of-our-brains/. Accessed May 5, 2011.

Dahlstrom, Eden, de Boor, Tom, Grunwald, Peter, and Vockley, Martha. *National Study of Undergraduate Students and Information Technology, 2011*. Boulder, CO, EDUCAUSE Center for Applied Research, October 2011.

Fuller, Richard Buckminster. *Education Automation: Freeing the Scholar to Return to His Studies*. London, Southern Illinois Press/Carbondale & Simons, Inc., 1962.

Gabriel, Trip, and Richtel, Matt. "Inflating the Software Report Card," published October 8, 2011, http://www.nytimes.com/2011/10/09/technology/a-classroom-software-boom-but-mixed-results-despite-the-hype.html?_r=1. Accessed October 9, 2011.

Gladwell, Malcolm. *Blink: The Power of Thinking Without Thinking*. New York, Little, Brown and Company, 2005.

Godin, Seth. *Linchpin: Are You Indispensable?* New York, Portfolio/Penguin Group, 2010.

Hall, Edward T. *The Hidden Dimension*. New York, Anchor Books, 1966.

Hurson, Tim. *Think Better: An Innovator's Guide to Productive Thinking*. New York, McGraw Hill, 2008.

McClean, Shilo T. *Digital Storytelling: The Narrative Power of Visual Effects*. Cambridge, MA, The MIT Press, 2007.

McLuhan, Marshall. *The Medium Is the Massage: An Inventory of Effects*. New York, Bantam Books, 1967.

Moran, Mike, Seaman, Jeff, and Tinti-Kane, Hester. *Teaching, Learning, and Sharing: How Today's Higher Education Faculty Use Social Media*. New York, Pearson Learning Solutions, 2011.

Osteen, Mark. *Don DeLillo—White Noise: Text and Criticism.* New York, Penguin Books, 1998.

Parker, Kim, Lenhart, Amanda, and Moore, Kathleen. "The Digital Revolution and Higher Education," published August 28, 2011, http://www.pewsocialtrends.org/2011/08/28/the-digital-revolution-and-higher-education/2/. Accessed October 28, 2011.

Piaget, Jean. "Piaget Rediscovered," *Journal of Research in Science Teaching.* Eleanor Duckworth, 2 (1964): 175.

Postman, Neil. *Amusing Ourselves to Death: Public Discourse in the Age of Show Business.* New York, Penguin Books, 1985.

Richtel, Matt. "Attached to Technology and Paying a Price," published June 6, 2010, http://www.nytimes.com/2010/06/07/technology/07brain.html?ref=yourbrainoncomputers&pagewanted=2. Accessed September 18, 2011.

——— "Growing Up Digital, Wired for Distraction," published November 21, 2010, http://www.nytimes.com/2010/11/21/technology/21brain.html?ref=yourbrainoncomputers. Accessed November 3, 2011.

——— "Multitasking Takes Toll on Memory, Study Finds," published April 11, 2011, http://bits.blogs.nytimes.com/2011/04/11/multitasking-takes-toll-on-memory-study-finds/?scp=1&sq=multitasking&st=cse. Accessed November 2, 2011.

——— "Silicon Valley Wows Educators, and Woos Them," published November 4, 2011, http://www.nytimes.com/2011/11/05/technology/apple-woos-educators-with-trips-to-silicon-valley.html?hp. Accessed November 5, 2011.

Rideout, Victoria, with Saphir, Melissa (Data Support). *Zero to Eight: Children's Media Use in America.* San Francisco, Common Sense Media, October 25, 2011.

Rushkoff, Douglas. *Program or Be Programmed: Ten Commandments of the Digital Age.* New York, OR Books, 2010.

Weiss, Scott. *Handheld Usability.* New York, John Wiley & Sons, 2002.

Studies in the Postmodern Theory of Education

General Editor
Shirley R. Steinberg

Counterpoints publishes the most compelling and imaginative books being written in education today. Grounded on the theoretical advances in criticalism, feminism, and postmodernism in the last two decades of the twentieth century, Counterpoints engages the meaning of these innovations in various forms of educational expression. Committed to the proposition that theoretical literature should be accessible to a variety of audiences, the series insists that its authors avoid esoteric and jargonistic languages that transform educational scholarship into an elite discourse for the initiated. Scholarly work matters only to the degree it affects consciousness and practice at multiple sites. Counterpoints' editorial policy is based on these principles and the ability of scholars to break new ground, to open new conversations, to go where educators have never gone before.

For additional information about this series or for the submission of manuscripts, please contact:

Shirley R. Steinberg
c/o Peter Lang Publishing, Inc.
29 Broadway, 18th floor
New York, New York 10006